FM 31-21

DEPARTMENT OF THE ARMY FIELD MANUAL

GUERRILLA WARFARE AND SPECIAL FORCES OPERATIONS

PrepperPress
Post-Apocalyptic Fiction & Survival Nonfiction

ISBN: 978-1-943544-27-1

HEADQUARTERS, DEPARTMENT OF THE ARMY

FIELD MANUAL No. 31-21

HEADQUARTERS,
DEPARTMENT OF THE ARMY
WASHINGTON 25, D.C., *29 September 1961*

GUERRILLA WARFARE AND SPECIAL FORCES OPERATIONS

* **This manual supersedes FM 31-21, 8 May 1958.**

PART ONE
INTRODUCTION

CHAPTER 1
FUNDAMENTALS

1. Purpose and Scope

a. This manual provides guidance in Special Forces and unconventional warfare operations for commanders and staffs at all levels. The basic concepts of unconventional warfare are presented in a manner designed to acquaint the reader with Special Forces organization, concepts, and methods of operations to fulfill the Army's responsibility for the conduct of unconventional warfare.

b. Thorough understanding of the ideas established within this manual will prepare the commander and staff officers for subsequent decisions and staff actions which affect special forces planning and operations.

c. Detailed methods and techniques of Special Forces operations are discussed below. Classified information pertaining to all levels of Special Forces operations is found in FM 31-21A.

2. Definition of Unconventional Warfare

Unconventional warfare consists of the interrelated fields of guerrilla warfare, evasion and escape, and subversion against hostile states (resistance). Unconventional warfare operations are conducted in enemy or enemy controlled territory by predominately indigenous personnel usually supported and directed in varying degrees by an external source.

3. Delineation of Responsibilities for Unconventional Warfare

a. The responsibility for certain of these activities has been delegated to the service having primary concern. Guerrilla warfare is the responsibility of the United States Army.

b. Within certain designated geographic areas—called guerrilla warfare operational areas—the United States Army is responsible for the conduct of all three interrelated fields of activity as they affect guerrilla warfare operations.

c. The military operations of resistance movements are customarily supported and accompanied by political and economic activities—both overt and clandestine—of individuals and groups integrated, or acting in conjunction with guerrillas. The several types of activities are interlocking. The term unconventional warfare is used in this manual to denote all of the United States Army's associated responsibilities in the conduct of guerrilla warfare. The term guerrilla warfare is used to denote the primary overt military activities of the guerrilla forces.

CHAPTER 2
RESISTANCE AND GUERRILLA WARFARE

4. Resistance

a. General. Resistance is the cornerstone of guerrilla warfare. Underground and guerrilla warfare stem from a resistance movement.

b. Definition. Resistance is defined as the act of opposition of one individual or group to another. A resistance movement is the organized element of a disaffected population which resists a government or occupying power with means, varying from passive to violently active. Resistance movements begin to form when dissatisfaction occurs among strongly motivated individuals who cannot further their cause by peaceful and legal means.

c. The Nature of Resistance.

(1) Resistance, rebellion or civil war begins in a nation where political, sociological, economic or religious division has occurred. Divisions of this nature are usually caused by a violation of rights or privileges, the oppression of one group by the dominant or occupying force, or the threat to the life and freedom of the populace. Resistance also may develop in a nation where the once welcomed liberators have failed to improve an intolerable social or economic situation. Resistance can also be deliberately inspired from external sources against an assumed grievance. Resistance can be active or passive. Passive resistance may be in the form of smoldering resentment which needs only leadership or a means of expression to mature to active resistance.

(2) Some people join a resistance movement because of an innate desire to survive. Others may join the resistance forces because of deep ideological convictions. But all, regardless of initial motivation, are bound together to fight against a common enemy. Part of the population assists the resistance movement as fighters in the guerrilla force; some assist as part-time guerrillas or in civilian support agencies know as auxiliary units; while others are members of the underground.

d. *Influencing Factors.*

(1) *Environment.*

(a) Terrain. The physical location of the resistance movement has a great influence upon its organization and tactics. Because they provide suitable areas for the security of operations, mountains, swamps, large forests or jungles nurture overt or guerrilla type resistance. Flat plains areas and large towns or cities are more apt to lead to underground resistance activities although the possibility of organizing a guerrilla force in these areas should not be overlooked.

(b) Cultural. A peoples' cultural environment also has its effects on resistance movements. The urge to bear arms, escape, and fight the enemy is dependent on the cultural background of the people. Men from rural or peasant environment, not subjected to tight governmental control, have more opportunity to show their hatred of the enemy occupation by overt and violent means such as guerrilla warfare. People from an industrialized and highly urbanized culture will resist with such activities as sabotage, propaganda, passive acts and espionage.

(c) Control of population. When an occupying power is able to exercise close and stringent control over the population, the resistance movement is conducted primarily in secrecy. When the police and military forces of the occupying power are diverted or otherwise ineffectual, the resistance movement may be conducted with primarily overt guerrilla actions.

(2) *Motivation.* Besides the geographical and cultural environment influencing guerrilla warfare, the sociological climate produces many motivating factors which have a profound effect upon the resistance movement. Strong individual motivation is essential to the formation of a resistance force. Although some individual motives are not ideal and, if openly expressed, may do harm to the guerrilla effort, the following are examples of what some of the true motives may be.

(a) Ideology. In guerrilla units some individuals have developed strong ideological motives for taking up arms. These ideologies take root in two broad areas politics and religion. The individual tends to subordinate his own personality to these ideologies and works constantly and solely for the "cause." In some resistance fighters, this motive is extremely strong.

(b) *Economic.* Many individuals join resistance movements to keep from starving or to keep from losing their livelihood. An organized resistance force may exert economic influence on individuals who fail to support their movement.

(c) *Personal gain.* Personal gain is the motivating force of some volunteers. An individual, so motivated, may change sides if he believes he can gain more by fighting for the opposing force.

(d) *Hate.* People who have lost loved ones due to enemy actions may fight against that enemy as a result of engendered hatred. Uncontrolled hatred can pose problems for the sponsor because it is difficult to curb the fanaticism of such individuals and properly direct their efforts.

(e) *Security.* If the resistance movement is strong or gives the impression of being powerful, many individuals join out of a feeling of personal safety. Usually, this situation occurs only after the resistance movement is well organized and the enemy has been weakened by other actions. Others join in order to escape recruitment into the service of the enemy.

(f) *Ego.* Personal motives such as power, pride, and adventure operate to some extent in all individuals. Depending upon the moral fiber of the individual, these motives may sustain him in times of great stress.

(g) *Fear.* Some individuals become a part of the resistance movement through no personal desire of their own. They join the movement out of fear of reprisals against themselves or their families.

(3) *Chance for success.* In addition to motivation and circumstances of environment, a population must feel that there is ultimately a chance for success or there can be no effective resistance movement developed. Active participation in any resistance movement is influenced by its chance for success.

(4) *Guidance.* Resistance movements stand or fall on the caliber of the leaders and other individuals in the organization. An understanding of the environmental and individual motivating factors will assist greatly those who desire to obtain the optimum from a guerrilla organization. An analysis of these factors plays an important part in evaluating potential resistance forces.

5. Guerrilla Warfare

Guerrilla warfare comprises combat operations conducted in enemy held territory by predominantly indigenous forces on a military or paramilitary basis to reduce the combat effectiveness, industrial capacity, and morale of the enemy. Guerilla operations are conducted by relatively small groups employing offensive tactics. Guerrilla warfare supports other military operations.

6. Characteristics of Guerrilla Warfare

a. General. Guerrilla warfare is characterized by offensive Guerrilalcatsiorne.ly upon mobility, elusiveness and surprise. In addition to these traits, there are other characteristics that should be mentioned: civilian support, outside sponsorship, political aspects, legal aspects, tactics, and development aspects.

b. Support Factors.

(1) *Civilian support.* The success of guerrilla movements depends upon continuous moral and material support from the civilian population. The local community usually is under intense pressure from anti-guerrilla factions. Punitive measures such as reprisals, terrorism, deportation, restriction of movement and seizure of goods and property are conducted against supporters of guerrilla activity, making this support dangerous and difficult. If the local populace has a strong will to resist, enemy reprisals cause an increase in underground activities. The civilian community may assist the guerrilla force by furnishing supplies, recruits, information; by giving early warning; by supporting evasion and escape; and other activities. After the guerrilla force has established itself and is sufficiently strong, it may need to exert force upon certain elements of the civilian population to command their support, e. g.: coerce indifferent or unresponsive portions of the population into supporting the guerrilla movement. Civilians participating in such support activities, aside from underground operations, comprise what are known as the auxiliary forces.

(2) *Outside sponsorship.* Guerrilla operations are more effective when outside sponsorship is present. During a wartime situation this support is political, psychological and logistical as well as tactical. A sponsoring power decides to support guerrilla forces when it feels that the guerrillas can make a significant contribution toward the achievement of national objectives.

c. Political Aspects.

(1) Guerrilla warfare has often been described as being more political than military in nature. It is certainly military in the tactical sense, but it is also political since a guerrilla movement generally stems from a local power struggle. Guerrillas usually fight for political gains, although in gaining their own political objectives they may assist the sponsoring power to gain its military objective.

(2) The political dominance in guerrilla warfare can be seen from another point of view. Guerrilla leaders with a common enemy, but politically opposed, may dissipate their efforts by fighting each other. The politically oriented guerrilla leader can cause trouble by withholding his cooperation until he extracts promises of political significance from his sponsor. The political imprint on guerrilla warfare is but another aspect that must be closely studied.

d. Legal Aspects. Guerrilla warfare is bound by the rules of the Geneva Conventions as much as is conventional warfare. As outlined in appropriate international agreements and FM 27-10, four important factors give a guerrilla legal status: (1) be commanded by a person responsible for the actions of his subordinates; (2) wear a fixed and distinctive insignia or sign recognizable at a distance; (3) conduct operations in accordance with the laws and customs of war; and (4) carry arms openly. If these four factors ire present, the guerrilla is entitled to the same treatment from his captors as the regular soldier. During World War II, General Eisenhower sent a proclamation to Nazis and Frenchmen alike, formally recognizing the French Resistance Maquis as members of the Allied Forces, and warned the Germans that all guerrillas were to be given the same honorable treatment as the regular soldiers under him in the Allied Expeditionary Force.

e. Tactics.

(1) *Primary considerations.* Guerrillas, because they are irregular soldiers, generally do not achieve unity of action in the same manner as conventional units. Because of this and two other factors—the logistical problem and manpower requirements—guerrillas initially cannot hope to meet and decisively defeat a conventional unit in a pitched battle. Guerrilla operations are facilitated by other military activities which distract potential enemy reinforcements. On the other hand, if the enemy is free of other concerns, he will combat the guerrillas with his best troops in order to protect vital installations. Guerrilla units, therefore, must coordinate their activities with other

friendly military forces and attack the enemy at points most disadvantageous to him. These attacks are normally conducted during periods of low visibility and are directed against isolated outposts, weakly defended locations or the moving enemy. By recognizing his own limitations and weaknesses, the guerrilla can hope for survival and eventual success. Initially, he is usually inferior to the enemy in firepower, manpower, communications, logistics, and organization. He is equal, and often superior, to the enemy in the collection of intelligence information, cover and deception, and the use of time.

(2) *Offensive tactics*. The basis of successful guerrilla combat is offensive action combined with surprise. During periods of low visibility, the guerrilla attacks, tries to gain a momentary advantage of firepower, executes his mission to capture or destroy personnel and equipment, and leaves the scene of action as rapidly as possible. Normally, the guerrilla does not consistently operate in one area but varies his operations so that no pattern is evident. If possible, he strikes two or three targets simultaneously to divide the enemy pursuit and reinforcement effort.

(3) *Defensive tactics*. Protective surveillance for the guerrilla is usually very good; he has civilian non-combatants providing him with information on enemy garrisons, troop movements, and counter, guerrilla activities. This advance warning gives the guerrilla time for proper countermeasures. If, in any counter guerrilla move by a superior enemy, the guerrillas are threatened or encircled, they do not meet him on a showdown basis, but withdraw, disperse or attempt a breakout.

f. Development Aspects. To complete the picture of guerrilla warfare, a time-and-space frame of reference must be understood. That is, it is not sufficient merely to state certain principles of guerrilla warfare, but it is necessary to qualify statements regarding guerrilla actions to fix them with regard to time and space.

(1) *Time element*. Guerrillas have proved themselves effective during all stages of conflict from the outbreak of hostilities until the end of fighting. However, in the early stages of guerrilla development, when the enemy is still strong, resistance operations normally tend to be conducted less openly. During this period, security is a prime concern. If the resistance movement is to survive and develop—while surrounded by strong enemy forces security is a prime concern and precautions must be extensive and effective. Activity is generally limited to information—gathering,

recruiting, training, organization, and small-scale operations.

(2) *Situation.* On the other hand, when the situation changes to favor the guerrillas either through enemy weakness or resistance-created favorable circumstances, operations become more overt making large-scale actions possible. When the situation permits, guerrilla forces expand and tend to adopt conventional organizations.

(3) *Location.* Guerrilla warfare takes on different aspects according to its geographic location. In some areas of the world guerrilla warfare has preceded the entry of regular troops; while in other areas, guerrilla movements have come into existence after the formal entry of regular troops. Additionally, in some areas the complex social structure and economic organizations are cogs in a vast system that is relatively easy to disrupt. The higher the degree to which a country has evolved industrially the more vulnerable it is to activities of the guerrillas. In less industrialized areas of the world, these complexities do not exist. The people are less dependent on one another for goods and services; disruption of one community does not necessarily cause extreme hardship in another. Thus, the impact of guerrilla warfare upon the population is not as great and guerrilla fighting is likely to be more prevalent. In judging • the potential for, and effects of, guerrilla warfare location is an important consideration.

7. Special Forces Operations

The value of coordinating guerrilla activities with conventional military operations and the need for peacetime planning and training by the potential sponsor have been recognized by the United States. The unit organized and trained to implement the Army's responsibility in directing guerrilla operations is the Airborne Special Forces Group. Special forces units may be called upon to operate during a general, limited or cold war.

a. General War. The doctrine set forth in this manual is structured around a general war situation. In a general war, Special Forces organize guerrilla forces to support conventional military operations under the direction of the theater commander. Their operations generally are conducted in denied (enemy controlled) territory.

b. Limited War. Limited war operations by Special Forces could be of the general type with infiltration of Special Forces detachments or of a training nature conducted in a nondenied area with infiltration of indigenous units only.

c. Cold War. Special forces units can assist in training military personnel in combating guerrilla and terrorist activities and subversion. In addition, they may train foreign military personnel in the techniques of guerrilla warfare, thus enhancing the defense capability of the nation concerned. When so employed, Special Forces units supplement the U.S. military assistance groups and army missions.

8. Capabilities and Limitations

a. Capabilities. Special forces deployment gives reach to the theater commander's operations. It permits him to influence activities far in advance of the field forces and beyond the range of army-controlled weapons systems. Special forces directed guerrilla units (called UW forces) conduct operations which are categorized as follows:

(1) *Missions in support of theater commander.* These missions include—

(a) Interdiction of lines of communications, key areas and military and industrial installations.

(b) Psychological operations.

(c) Special intelligence tasks such as target acquisition and damage assessment.

(d) Evasion and escape operations.

(e) Cover and deception operations.

(2) Missions to support combat operations of tactical commanders. In addition to an intensification of the tasks listed in (1) above, UW forces execute missions to directly assist conventional forces engaged in combat operations. Such missions may include—

(a) Seizure of key terrain to facilitate airborne and amphibious operations.

(b) Employment as a reconnaissance and security force.

(c) Seizure of key installations to prevent destruction by the enemy.

(d) Diversionary attacks against enemy forces to support cover and deception plans.

(e) Operations which isolate selected portions of the battle area, airborne objective area or beachhead.

(3) Missions conducted after juncture with friendly forces. In the event control of guerrilla units is retained by the United States, the following missions may be assigned:

(a) Reconnaissance and security missions.

(*b*) When properly trained and supported, conventional combat operations.

(*c*) Rear area security missions.

(*d*) Counter-guerrilla operations.

(*e*) Support of civil affairs operations.

b. Limitations. It must be realized that there are certain limitations in the use of guerrilla forces. Some of these limitations are

(1) Limited capabilities for static defensive or holding operations.

(2) Initially, lack of formal training, equipment, weapons, and supplies prohibit large-scale combat operations.

(3) Dependence upon the local civilian population and an outside sponsoring power for supplies and equipment.

(4) Communications between the guerrilla warfare operational area and higher headquarters in friendly territory are often tenuous and fraught with technical problems.

(5) Decentralization of command and dispersion of forces for security impedes reaction time to orders from theater level.

(6) Restrictions on friendly supporting fires into the operational area because of necessity for frequent moves by the guerrillas as well as the necessity far protecting the friendly civilian population so far as possible.

(7) From initial contact until an operation is completed, the entire project is dependent upon precise, timely and accurate intelligence.

PART TWO
ORGANIZATION FOR THE SPECIAL FORCES EFFORT

CHAPTER 3
JOINT UNCONVENTIONAL WARFARE TASK FORCE (JUWTF)

9. General

a. The theater commander is responsible for the conduct of unconventional warfare in his area of operations.

b. As a part of this responsibility he designates guerrilla warfare operational areas for the conduct of guerrilla warfare and related unconventional warfare activities.

10. Organization of the Joint Unconventional Warfare Task Force (JUWTF)

a. The theater commander has the authority to organize his command for unconventional warfare in the manner best suited to accomplish his mission.

b. The preferred organization is a joint subordinate headquarters for unconventional warfare on the same level as other service component commands (fig. 1) .

This subordinate headquarters, known as a Joint Unconventional Warfare Task Force (JUWTF), is composed of representatives from the service component commands and appropriate civilian personnel.

c. *A second possible organization is an unconventional warfare plans section within the J3 staff division of the unified or specified command.

d. The internal staff organization of the JUWTF is joint, with the principal staff officers being from any service, and consisting of a J1, J2, J3, J4, J6 divisions and any required special staff officers (fig. 2). In the JUWTF the plans and policy functions of J5 division are normally accomplished by the J3 division.

e. Units and individuals from the service components are assigned or attached for operational control to the JUWTF.

* Since an unconventional warfare plans section within the JS division performs essentially the same functions as a separate JUWTF, further discussion is limited to the separate JUWTF.

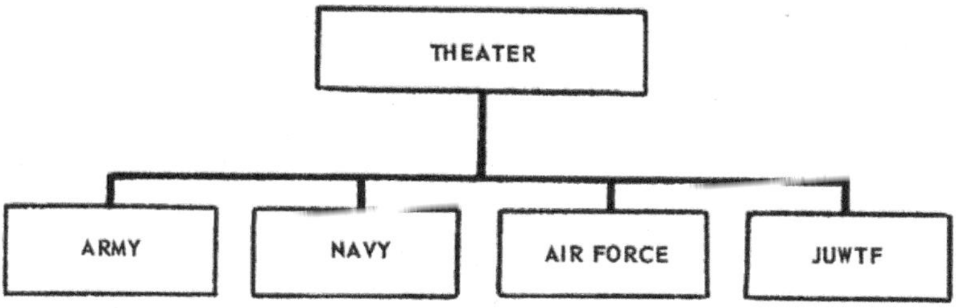

Figure 1. A theater organization.

f. The airborne special forces group, the principal army element of the JUWTF, establishes a special forces operational base (SFOB) to command and support operational detachments before and after commitment in designated guerrilla warfare operational areas.

11. Functions of the Joint Unconventional Warfare Task Force

a. The JUWTF commander and his staff make operational plans for and direct the conduct of unconventional warfare. The principal functions of the JUWTF are

(1) Recommend geographical areas to be designated guerrilla-warfare operational areas.

(2) Procure and maintain intelligence materials in support of unconventional warfare.

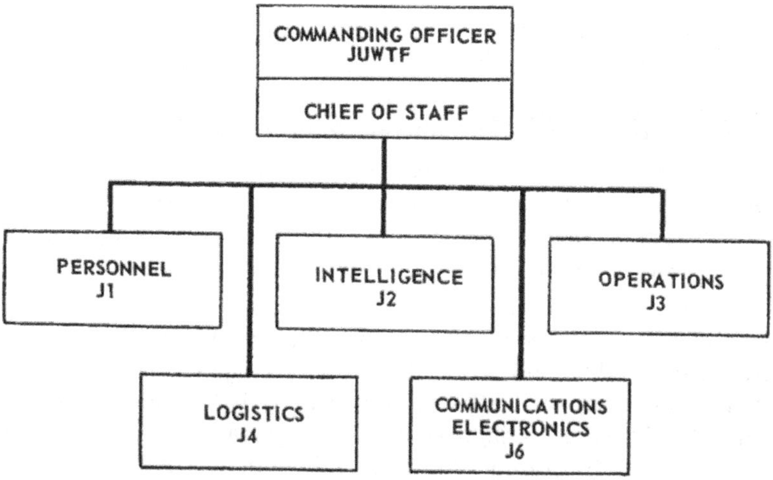

Figure 2. JUWTF organization.

(3) Develop operational, administrative and logistical plans and requirements for the support of unconventional warfare.

(4) Coordinate with other theater agencies in planning for all types of operations.

(5) Develop communication procedures and requirements to support unconventional warfare plans.

(6) Plan and conduct joint training of land, sea and air units designated to participate in or support unconventional warfare.

(7) As directed, coordinate with allied military authorities for the preparation and execution of unconventional warfare plans.

(8) Maintain liaison with other unconventional warfare agencies or units.

(9) Recommend strengths of indigenous forces to be supported for unconventional warfare operations.

(10) Maintain liaison at staff and operational level with appropriate intelligence agencies; coordinate requirements, collection and communications with other activities in denied areas; plan intelligence operations in support of conventional forces when directed by the theater commander.

(11) Maintain liaison with theater civil affairs units with respect to civil affairs (CA) aspects of unconventional warfare.

b. The staff operations of a JUWTF are basically the same as for other US military staffs.

12. Operational Control of Unconventional Warfare Forces

a. Initially, operational control of US sponsored unconventional warfare forces is retained by the theater commander. Control is exercised through the JUWTF assigning missions to the Special Forces group, which in turn directs deployed operational detachments.

b. When guerrilla warfare operational areas fall within the area of influence of advancing tactical commands, operational control of affected unconventional warfare forces usually is transferred from the unified or specified command level through theater army to the field army concerned. In conjunction with this transfer, elements of the Special Forces group are attached to the army headquarters to provide continuity of supervision.

c. The field army commander in turn may assign operational control of the unconventional warfare force to any of his subordinate tactical units. Delegation of control generally is not made lower than division. See chapter 8 for a more detailed discussion of utilization of unconventional warfare forces by tactical commands.

CHAPTER 4
AIRBORNE SPECIAL FORCES GROUP

Section I. GENERAL

13. General

The Airborne Special Forces Group is the United States Army's organization trained to conduct guerrilla warfare and related unconventional warfare activities. Special forces are a strategic force employed under the direction of theater commanders. Deployment of special, forces units allows the theater commander to conduct offensive operations deep in enemy territory.

14. Mission and Concept

a. Mission. The mission of special forces is to develop, organize, equip, train, and direct indigenous forces in the conduct of guerrilla warfare. Special forces may also advise, train and assist indigenous forces in counter-insurgency operations.

b. Concept. Special forces are responsible for the conduct of all unconventional warfare activities within guerrilla warfare operational areas and may be called upon to perform other tasks associated with or in support of guerrilla warfare.

15. Airborne Special Forces Group

a. Organization. The Airborne Special Forces Group consists of a headquarters and headquarters company and four Special Forces companies (fig. 3) .

b. Capabilities. The special forces group is capable of—

(1) Deploying its operational detachments by air, sea or land when provided with appropriate transportation.

(2) Organizing, training, and directing a number of varied size guerrilla units.

(3) Controlling, by long-range communications, the operations of UW forces in enemy or enemy occupied territory to reduce his combat effectiveness, industrial capacity, and morale.

(4) Performing specialized intelligence missions as directed by higher commanders and when augmented by intelligence specialists as required.

(5) Providing training and assistance to friendly foreign armies in guerrilla and counter guerrilla operations.

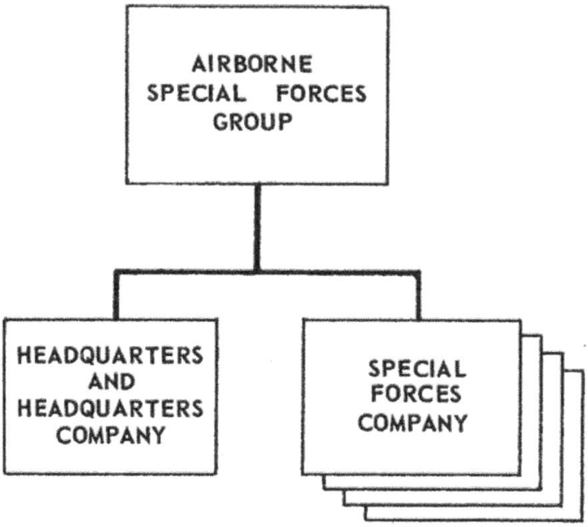

Figure 3. Airborne special forces group.

(6) Establishing a Special Forces operational base when augmented by support and service units.

c. Additional Considerations.

(1) The Special Forces group requires augmentation by support and service units to conduct sustained operations from the Special Forces Operational Base (SFOB). For details of the support required, see paragraph 21.

(2) The reaction time of Special Forces detachments differs from that of conventional infantry units because of communications limitations and greater distances to operational areas.

16. Headquarters and Headquarters Company (fig. 4)

a. Mission. To provide communications, administrative, training, intelligence, and logistical support for assigned Special Forces elements prior to and after deployment.

b. Capabilities. Headquarters and headquarters company of the special forces group has the following capabilities:

(1) Provides command and staff control and planning for Special Forces elements prior to and after deployment.

(2) Provides logistical support (except delivery) to Special Forces operational elements on a continuing basis.

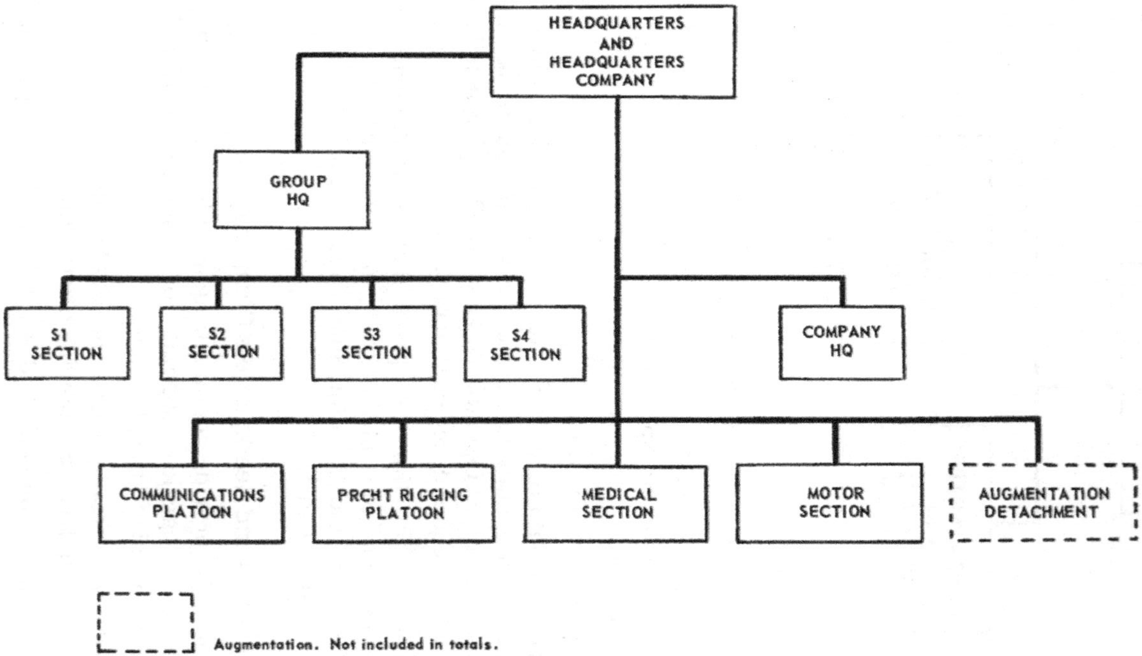

Figure 4. HQ and HQ Co, airborne special forces group.

(3) Provides long-range communication between an SFOB and assigned Special Forces operational elements.

(4) Provides unit level medical and dental service.

(5) Provides third echelon maintenance in radio and small arms.

c. Organization. For detailed organization, see the current tables of organization and equipment.

17. Special Forces Company (fig. 5)

a. Organization. The special forces company consists of an administrative detachment, one operational detachment C, three operational detachments B, and 12 operational detachments A.

b. Administrative Detachment. The administrative detachment performs the normal administrative functions of a company head-

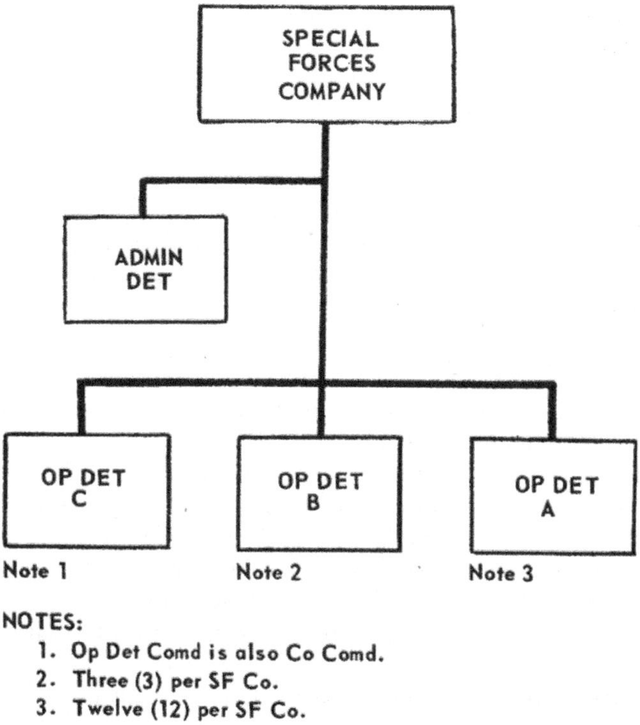

Figure 5. Special forces company.

quarters. The detachment commander executes the directives and orders of the commanding officer of the operational detachment C, who is the commander of the Special Forces company. During operations, the administrative detachment reverts to the group commander's control when its operational detachments have been deployed.

c. Operational Detachment C (fig. 6). Operational detachment C is the senior operational unit of the Special Forces company. It is capable of—

(1) Conducting operations with guerrilla forces.

(2) Exercising operational control over designated subordinate detachments.

(3) Providing the nucleus of a liaison detachment to field army or other tactical elements when the operational control of Special Forces detachments is passed to tactical unit commanders. When so employed, the detachment C is attached to the tactical unit headquarters.

d. Operational Detachment B (fig. 7) . Operational detachment B, like the detachment C, conducts operations with guerrilla forces. When operating with other detachments, the B detachment exercises operational control over subordinate detachments and/or is subordinated to a senior detachment. It also can provide the nucleus of a liaison detachment at a tactical unit headquarters.

e. Operational Detachment A (fig. 8). The operational detachment A conducts operations with guerrilla forces, either unilaterally or in conjunction with other detachments. When operating with other detachments, the A detachment is normally subordinated to a senior detachment.

18. Augmentation Detachment

a. The augmentation detachment is identical in composition to the operational detachment C (fig. 6).

b. The augmentation detachment is capable of:

(1) Assisting the commander in the control of operations.

(2) Forming the liaison detachment at a tactical unit headquarters when operational control of Special Forces detachments is passed to the tactical unit.

(3) Forming the nucleus to establish an alternate SFOB headquarters.

(4) Conducting operations in guerrilla warfare operational areas.

c. For a detailed description of the employment of the augmentation detachment, see paragraph 21.

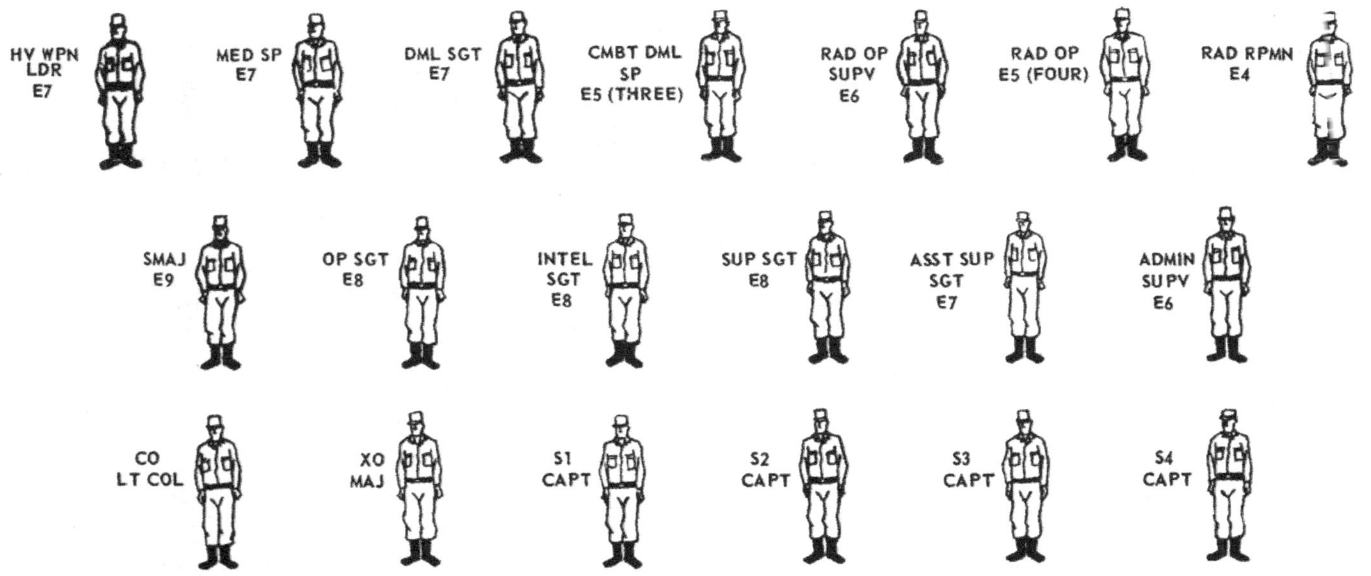

Figure 6. Operational detachment C.

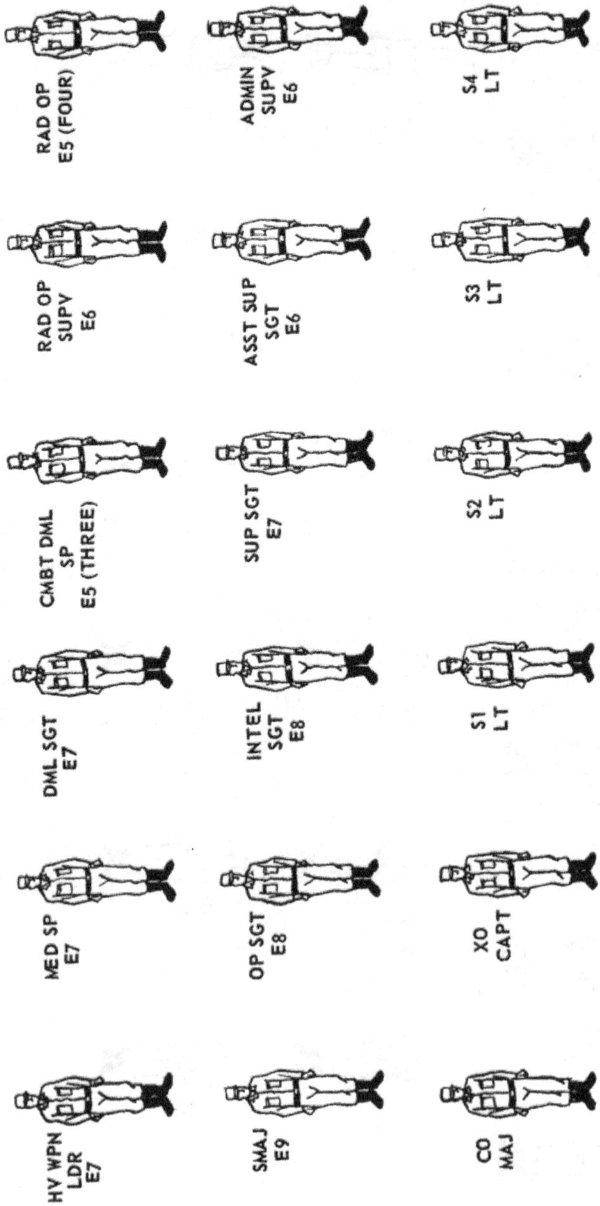

Figure 7. Operational detachment B.

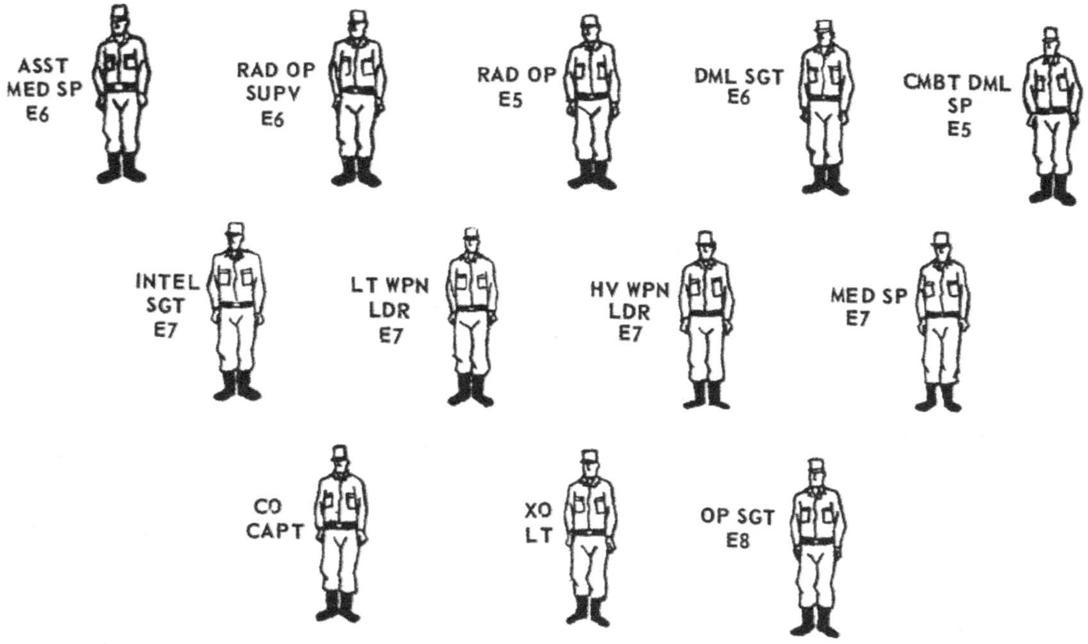

Figure 8. Operational detachment A.

Section II. THE SPECIAL FORCES OPERATIONAL BASE

19. General

a. The Special Forces operational base (SFOB) is the operational and administrative focal point for guerrilla warfare activities within a U.S. theater of operations. It is located in territory under friendly control, usually within the communications zone (CommZ). The location of the SFOB is designated by the theater commander.

b. The Special Forces group, supported by elements within the CommZ, activates and operates the SFOB. Ideally, the SFOB is established and activated prior to the commencement of hostilities. However, since preemergency activation may not be authorized, the base often is established on a standby basis to include long-range communication facilities, intratheater and intrabase communications, and emergency stockpiles of equipment.

c. The SFOB may be physically located at one installation or dispersed among a number of small sites, usually in the vicinity of other installations such as an air base or CommZ depot. Various modifications are adopted to suit the local situation.

20. Mission

The mission of the SF'OB is to prepare operational detachments for deployment into guerrilla warfare areas and, after deployment, to direct, administer, and support guerrilla forces in furtherance of the theater mission.

The functions performed at the SFOB are—

a. Planning and direction of operations.

b. Communications support.

c. Intelligence support.

d. Logistical support.

e. Briefing and staging.

f. Infiltration.

g. Liaison and coordination.

h. Training.

i. Administration.

21. Organization

(fig. 9)

a. *General.* The SFOB is organized along functional lines into two major groups: the operational element.,; and the administrative and training elements.

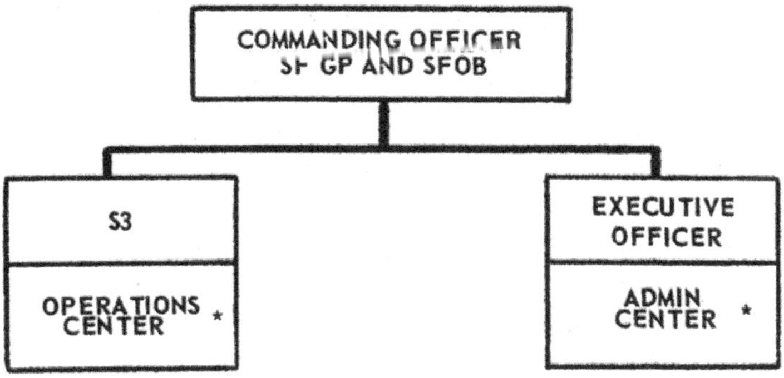

*Staff representation included in both operations and administrative centers.

Figure 9. Special forces operational base.

b. Operational Elements. The operational elements of the SFOB consist of—

(1) *Operations center.* The operations center is the facility within which are grouped representatives of unit and special staff sections and other commands concerned with current operations in guerrilla warfare areas. For a detailed discussion, see paragraph 23.

(2) *Briefing center.* The briefing center is an isolation area or areas where Special Forces detachments receive their operational missions and conduct final preparation for infiltration. For a detailed discussion, see paragraph 24.

(3) *Communications complex.* The communications complex consists of the communications facilities available to support the SFOB and guerrilla warfare operational areas. It includes the Special Forces group communications platoon, plus attached or supporting signal units and facilities. The group signal officer is the staff supervisor. The communications platoon leader is normally the operator and supervises the signal installations. For a detailed discussion, see paragraph 25.

(4) *The Augmentation detachment.*

(*a*) When activated, the augmentation detachment provides additional flexibility to the Special Forces group commander.

(*b*) The augmentation detachment may be employed to assist the commander in the control of operations. When so employed the detachment commander becomes a deputy for operations and supervises the operational elements of the SFOB. Other detachment personnel are assigned duties to operate the briefing center and supplement staff elements of the base.

(*c*) The detachment may form the Special Forces liaison detachment with a field army or other tactical command. In this situation the detachment is attached to the tactical command and exercises operational control for the tactical commander over designated guerrilla warfare operation areas (ch. 8) .

(*d*) The detachment can establish an alternate SFOB headquarters. In this role additional personnel and equipment are provided from headquarters and headquarters company and other sources as required.

(*e*) The detachment can be employed operationally to accomplish tasks appropriate for a C detachment.

(*f*) More than one augmentation detachment may be activated for employment at the SFOB.

c. Administrative and Training Elements. The administrative and training elements of the SFOB consist of—

(1) *Administrative center.* The administrative center is the facility within which are grouped representatives of unit and special staff sections and other commands concerned with current administrative support operations, base security, and area damage control. For further discussions, see paragraph 29.

(2) *Logistics support element.* This is a non-TOE grouping of Special Forces and supporting technical service units from CommZ formed to support the SFOB and guerrilla warfare operational areas. The group S4 supervises operations of the logistics support element which includes

(*a*) Organic elements of the Special Forces group: Supply Section, Motor Section, Parachute Rigging Platoon, and the Medical Section.

(*b*) Supporting elements as required: Transportation units, Engineer Utility Personnel, Ordnance 3d Echelon Support, Civilian Labor, QM Aerial Supply Units, Medical Units, and a Liaison Section from CommZ.

(3) *Other supporting units.* Although the group has personnel to establish the SFOB, deploy detachments and provide limited logistical support, it requires augmentation to conduct support activities on a sustained basis. CIC, base security and logistical support elements are required to support initial operations. As the number of operational areas increases with the subsequent buildup in guerrilla forces, the administrative support operations expand correspondingly. Units, such as those outlined in paragraph (2) above, are required to augment the SFOB. Military police security units and counterintelligence corps teams are included and operate under the headquarters commandant and S2, respectively. An army aviation detachment may be attached to the SFOB to provide army aviation support. An Army Security Agency unit may monitor communications for security. In some instances technical service units or installations are not located at the SFOB but provide general or direct support as a part of their mission. In this situation, the SFOB exercises no operational control over the units concerned but is serviced as a "customer" of the supporting unit or installation. An example is higher-echelon ordnance and engineer support.

(4) *Headquarters and headquarters company.* The headquarters and headquarters company, augmented by technical service and security units from CommZ, handles housekeeping activities at the SFOB. The company commander is the headquarters commandant.

(5) *Special forces companies (uncommitted units).* The uncommitted companies and detachments continue unit preparation and training. These detachments are briefed frequently on the situation in their projected operational areas.

Section III. CONTROL OF OPERATIONS

22. General

The Special Forces group organizes functionally to control operations in guerrilla warfare operational areas. The elements used in the control of operations are—

(1) Operation center.

(2) Briefing center.

(3) Communications complex.

23. Operations Center

a. General. The operations center is a functional grouping of TOE personnel who coordinate and control operations for the commander of the SFOB.

b. Functions.

(1) Detailed planning for guerrilla warfare operational areas, to include preparation of the operation plan for each operational detachment. This planning is based upon the UW plans of the theater commander.

(2) Conducting briefings and supervising other preparation by detachments assigned to the briefing center.

(3) Coordinating with other services and agencies as necessary.

(4) Exercising operational supervision over guerrilla war fare operational areas.

(5) Making recommendations concerning employment of guerrilla forces in support of military operations. This includes reorganization as necessary.

(6) Acting as the control and coordinating center for guerrilla warfare operational areas.

c. Composition (fig. 10).

(1) S3. The S3 exercises primary staff responsibility for operations center and is the director.

(2) *Plans element.* The operations center plans element conducts planning for future operations. The plans element consists of the assistant S2 and assistant S3 plus enlisted augmentation. The assistant S3 is the officer-in-charge of plans element. Once plans are approved they are implemented by the appropriate area specialist team.

(3) *S2 operations element.* The S2 operations element consists of the S2, intelligence sergeant, intelligence editors, analysts, and order-of-battle specialists. They assemble and evaluate intelligence information received from the operational areas; prepare and disseminate intelligence reports based on evaluated information and intelligence reports from other headquarters; and conduct intelligence briefings and debriefings. For a detailed discussion, see paragraph 27.

(4) *Assistant S4 (plans).* The Assistant .S4 (Plans) is the logistical coordinator for the Operations Center. He processes logistical requirements from the area specialist teams and is responsible for logistical activities in the

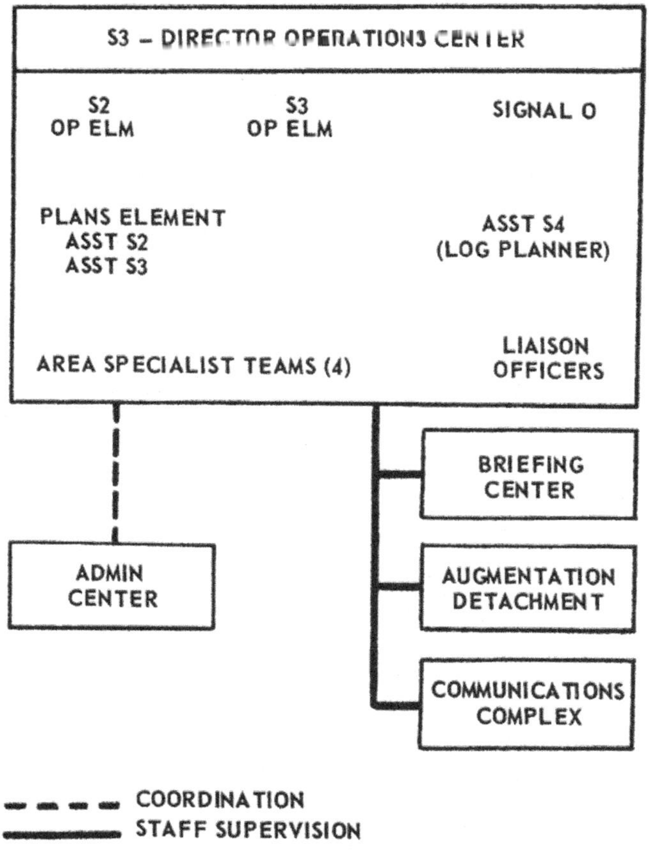

Figure 10. Staff relationships, the operations center.

briefing center. He prepares the administrative annex to the guerrilla warfare area operation plans.

(5) *Signal officer*. The signal officer is signal coordinator for the operations center. He coordinates signal requirements from area specialist teams and is responsible for operational signal matters. He prepares the signal operating instructions and signal annex for guerrilla warfare area operation plans.

(6) *Area specialist teams (ASTs)*. The ASTs are the focal point of the operations center. They assist in precommitment planning, coordinate activities of their assigned detachments in the briefing center, and act as the parents of the committed detachments. The AST consists of the area

specialist officer (assistant S3) and an area supervisor (senior NCO). This team must become expert on the specific area or country it will supervise during operations. The AST acts as the committed detachment's rear headquarters, and is responsible for following through on all directives to and messages from committed detachments. During preinfiltration briefings, a close rapport is established between the detachment and the AST. The AST keeps the commander and staff informed on the operational situation.

(7) *Communication center.* The communications center, operated by the command operations center team of the communications platoon, is located in the vicinity of the operations center since it provides the communications center support for the operations center and other elements of the SFOB.

(8) *Liaison officers.* Although not an integral part of the operations center, the liaison officers from the various services, field armies, and allied countries are located there. They coordinate matters of common interest with their services and nations and arrange support when required. They keep the Special Forces group commander and staff abreast of the situation of their respective organizations as these organizations influence guerrilla warfare areas of interest.

24. Briefing Center

a. General. The briefing center provides for the following:

(1) Housing.

(2) Messing.

(3) Briefing and debriefing.

(4) Detachment study.

(5) Dispensary service.

(6) Special training.

(7) Storage and packaging of accompanying supplies.

(8) Limited morale services.

(9) Staging of detachments to departure sites.

b. Operation. The entire area (areas) is a maximum-security site accessible only to those personnel who have a requirement to be there. The operations of the center are supervised by the S3. The headquarters commandant is responsible for the administrative functioning and security of the area. Personnel from the augmentation detachment and Special Forces company administrative detachment operate the briefing center.

c. Functioning.

(1) The director of the operations center (S3) coordinates the briefing and staging activities of the center. He schedules briefings and arranges for the staging of the detachments to the departure installation. Briefing personnel are the area specialist officers and the staff officers from the operations center, augmented, when required, by other members of the group staff and appropriate liaison officers. Often, specialists from other headquarters, services and allied governments participate.

(2) Detachments prepare their own plans based upon the operation plan for the guerrilla warfare area. Detachments package their own equipment. Parachute rigging support is provided as necessary. Detachments are afforded maximum time to study the material received at briefings. Since much of this information is classified, it is committed to memory. Essentials which do not lend themselves to memorization are miniaturized.

(3) Debriefings are conducted in the same manner as briefings, with recovered detachments remaining in isolation in the briefing center until the debriefing is complete.

(4) Detachment training conducted while in the briefing center is limited to that essential for the operation which could not be conducted elsewhere. New items of equipment or weapons issued at the last minute require familiarization or test firing. Specific techniques relating to infiltration may have to be taught. Identification and recognition of new or specific items of enemy equipment may be a requirement. If detachments are isolated for relatively long periods of time, training programs are expanded to maintain basic skills and physical fitness.

(5) Hospitalization of sick or injured members of detachments preparing for commitment is done so far as possible within the briefing center. A small dispensary facility is established to care for those personnel whose illness is not serious enough to preclude participation in their detachment's operational mission. The seriously sick or injured are evacuated to CommZ medical installations.

Arrangements are made to isolate sensitive personnel who are hospitalized outside the briefing center.

(6) The staging of detachments in the departure installation is arranged by the S3. The ASTs are directly responsible for the operation and accompany their detachments from the briefing center to the departure installation. The requirements of the delivery agency pertaining to preflight or embarkation briefings are coordinated in advance. If necessary, arrangements are made for secure housing at the departure installation.

25. SFOB Communications Complex

a. Organization of the Communications Platoon (figs. 11 and 12).

(1) The communications platoon headquarters provides the group signal officer with the necessary administrative and supply personnel for the operations of the platoon.

(2) The command operations center team operates the communications center serving the SFOB. The forward operations center team provides a command and control

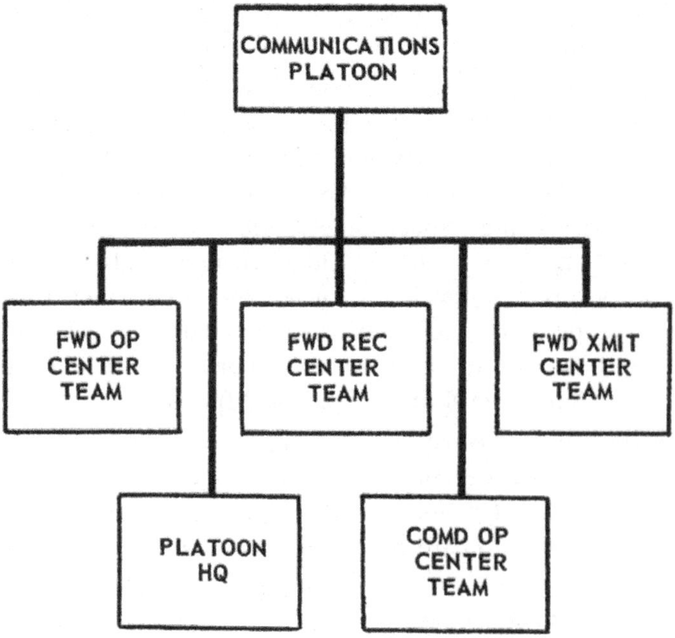

Figure 11. Communications platoon.

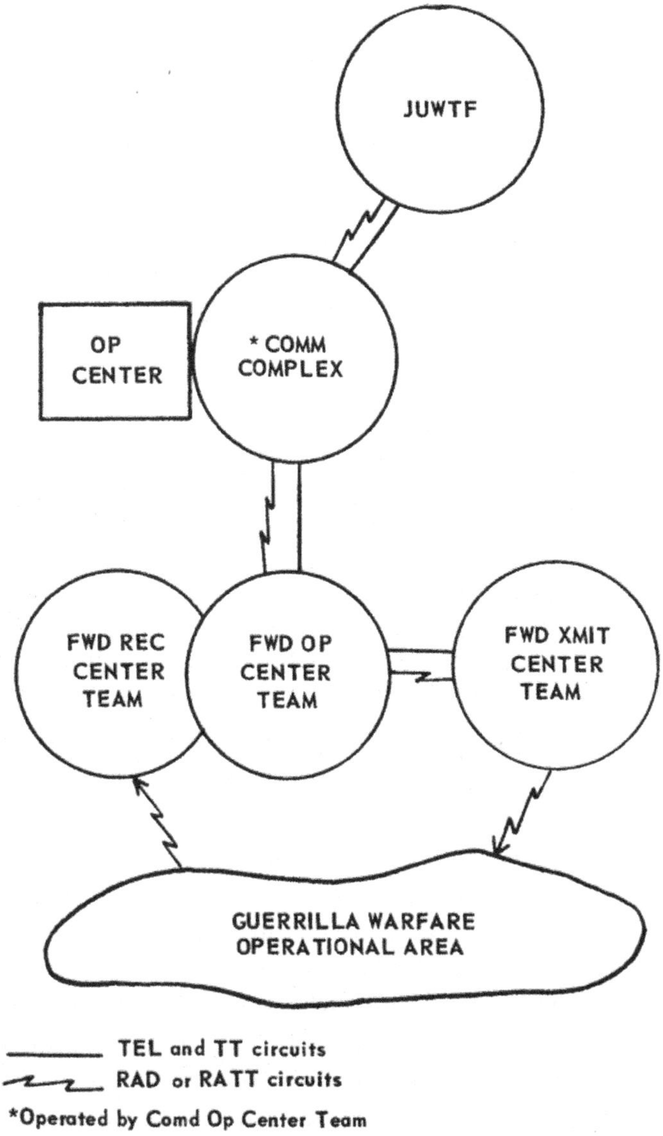

Figure 12. SFOB communication complex.

facility for the receiver and transmitter sites. The forward receiving center and the forward transmitting center teams operate the receiver and transmitter sites, respectively.

b. Capabilities.

(1) Operate telephone service within the SFOB. The platoon installs the telephones to be used within the SFOB and operates a telephone switchboard.

(2) Terminate lines from other headquarters. The primary means of communication to JUWTF and other theater agencies within friendly territory is telephone and teletype. Theater army signal troops install these long wire lines. The communications platoon terminates these lines in teletype, on-line crypto, crypto, crypto speech equipment, or telephone.

(3) Operate base receiver and transmitter sites. The SFOB receiver and transmitter sites may be separated from the base by a considerable distance. The base proper is connected to the receiver and transmitter sites by land line, telephone and teletype. Such lines are provided by theater signal agencies.

(4) Provide radio teletype back-up. The communications platoon provides radio teletype back-up to the land lines between the SFOB and the receiver and transmitter sites.

(5) Operate the communications center. The communications platoon, encrypts and decrypts messages and acts as a central and clearing center for the remote transmitter and receiver sites.

c. Responsibilities of the signal officer. The group signal officer is responsible for—

(1) Determining operating frequencies and communication security measures necessary to insure reliable radio communications with operational detachments.

(2) Obtaining frequencies as necessary and coordinating the use of all frequencies with the supporting signal officer.

(3) Signal planning and publishing necessary SOIs and SSIs.

(4) Planning, requesting and coordinating the engineering assistance necessary for the design of antenna farms and keying lines.

(5) Requesting material necessary to install these antenna farms and keying lines.

(6) Requesting signal support necessary to install long and local wire lines, e.g., SFOB to installations of the communication complex.

(7) Requesting the telephone and teletype trunks necessary for the operation of the SFOB.

(8) Advising the group S4 on signal supply matters.

(9) Supervising training of the group communications section and when directed training of all communication personnel in the group.

d. Signal Planning.

(1) Advance planning is necessary to provide reliable communications between the SFOB and guerrilla warfare operational areas. The distances between the SFOB and detachments, the communication security problems presented by operating from within enemy territory, and the low-power communication equipment used by the detachments all present technical problems. Proper frequencies must be chosen arid the use of these frequencies coordinated with the theater signal officer if communications are to be reliable. For further discussion of communications considerations, see chapters 5 and 7.

(2) The group signal officer insures that the JUWTF signal officer is aware of all aspects of his problem to include the extent of signal support required.

(3) The planning must be done as far in advance as possible; the reliability of the communications will be directly proportional to prior planning and supervision.

26. Plans

Planning, particularly prior to commitment of operational detachments, is one of the major functions of the special forces group. The S3 plans group is the focal point of planning functions. The ASTs, supervised by the assistant .S3 (plans) and assisted by other staff officers, accomplish operational planning. Considering the location, the mission and the ultimate developmental objectives, the ASTs prepare the operation plans for the guerrilla warfare operational areas. Planning is continuous and plans are revised as required. Wide latitude for the operational detachment is the rule for guerrilla warfare operation plans. The selection of a detachment for a particular mission is based on several factors, chief of which are: training status of the detachment and the ability and experience of the detachment commander. For security reasons, detachments do not have access to operation plans until assigned to the briefing center. However, to facilitate area studies, geographical regions (which include the detachments' specific guerrilla warfare operational areas) are assigned in advance.

27. Intelligence Section

a. General. The intelligence section of the special forces group is responsible for the following:

(1) Intelligence training.

(2) Furnishing intelligence to detachments prior to and after commitment.

(3) Conducting intelligence briefings and debriefings.

(4) Field evaluation of intelligence information received from guerrilla warfare operational areas.

(5) Counterintelligence.

(6) Supervisory responsibilities in coordination with the S1 for the exterior and interior security of the operational base and implementation of a security-education program.

b. Organization and Functions.

(1) *S2.* The S2 supervises the activities of the intelligence section, keeps the commander and staff informed of the intelligence situation, and coordinates with other staff sections and agencies.

(2) *Administration and training.* Intelligence administration and training is under the supervision of the training officer. He plans and supervises intelligence training and conducts routine administration for the section such as—processing security clearances, handling classified documents, and distribution of intelligence material within the SFOB.

(3) *Intelligence planning.* The assistant S2 is the principal intelligence planner at the SFOB. He represents the S2 section in the tactical operations center plans element. He conducts intelligence planning, prepares the intelligence annexes for the operation plans, and coordinates with other personnel of the intelligence section for specific intelligence support required.

(4) *Intelligence operations.* The S2 directs the activities of the intelligence sergeants, intelligence editor and analysts and the order of battle specialists to provide the intelligence support to guerrilla warfare operational areas and other headquarters. These personnel prepare estimates, plans, and summaries, routine and special intelligence reports, process information received from and furnish intelligence to committed detachments, prepare and maintain order of battle files, coordinate intelligence matters with other units

and headquarters, and prepare and conduct briefings and debriefings.

c. Attached or Supporting Intelligence Elements. The special forces group operating from an SFOB requires additional military intelligence support than that which is organically available. The group has no counterintelligence capability and requires CIC augmentation. Additional military intelligence specialists may be attached to assist in briefing detachments.

Section IV. CONTROL OF ADMINISTRATIVE AND TRAINING ACTIVITIES

28. General

a. The special forces group establishes an administrative center at the SFOB to control administrative and training activities.

b. The special forces group executive officer supervises the administrative center and other elements located at the SFOB that are engaged in administrative and training activities.

29. Administrative Center

(fig. 13)

a. Composition. The administrative center consists of

(1) The executive officer who is the director.

(2) Group S1.

(3) Group S4.

(4) Training officer, S2 Section.

(5) Training officer, S3 Section.

(6) Enlisted specialists as required.

b. Functioning. The administrative center plans and controls administrative and training activity at the base and directs the various sections, units arid attached elements in execution of their support tasks. Through coordinated planning the administrative center insures that guerrilla warfare operational areas and the SFOB receive the administrative support they require and that uncommitted operational detachments are trained for their missions.

30. Training

Training at the SFOB is accomplished under two conditions that conducted prior to isolation in the briefing center and that conducted in the briefing center. Training prior to receipt of an operational mission is intended to keep the detachment at its peak, to teach

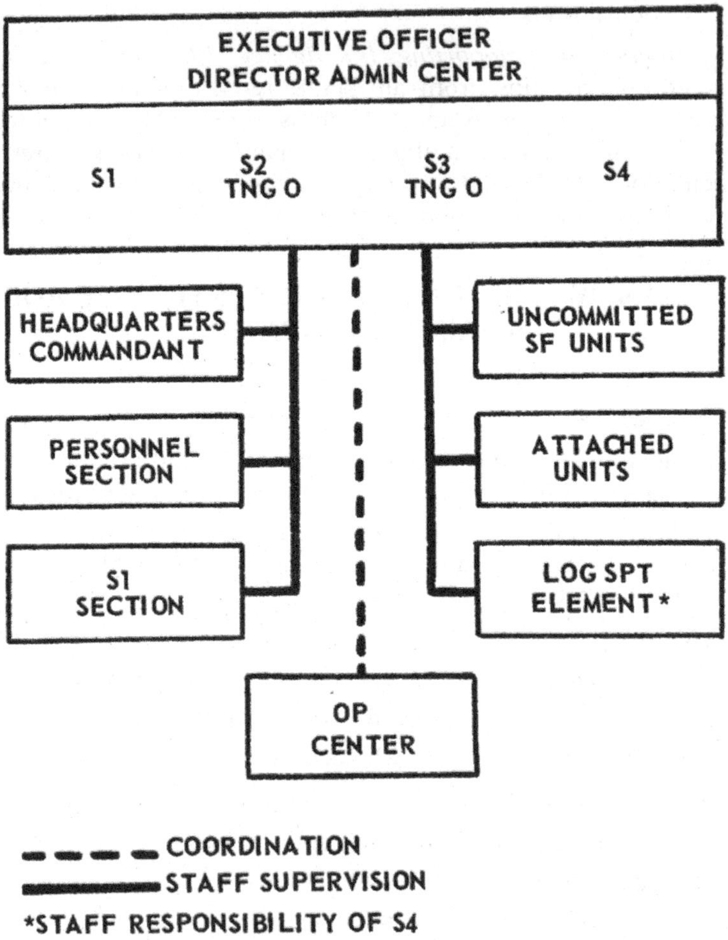

Figure 13. Staff relationships, administrative center.

specific techniques applicable to projected operations and to familiarize personnel with new equipment. Training conducted after assignment to the briefing center may include any or all of these, time permitting. Training areas include range facilities for test-firing and zeroing weapons and training with new equipment. In addition, plans are prepared to train replacements and/or replacement detachments. Training supervision is accomplished through the normal chain of command. The S3 training officer exercises staff supervision of training.

31. Administration

a. S1. The functions of any unit S1 are applicable to the special forces group S1; however, they must be modified to meet the situation which exists after deployment of operational detachments. Obviously the actions which are normal in other military units are difficult or impossible to accomplish when dealing with committed special forces personnel. The S1 prepares SOP's to cover foreseeable contingencies and takes steps to accomplish routine personnel matters prior to commitment. Personnel actions requiring a soldier's approval after he is in the operational area should be prepared in brevity codes to reduce radio transmissions. The S1 conducts portions of the predeployment briefing in the briefing center. Certain functions of the special forces group S1 are discussed wherein they are peculiar to deployed personnel.

(1) *Strengths.* Status of personnel is reported only when a change takes place, i.e. wounded, missing, captured, or killed.

(2) *Replacements.* The provisions of replacements depends upon the capability of the operational detachment to receive them and theater service components to deliver them. Replacements are provided on an individual or detachment basis.

(3) *Discipline, law and order.* Commanders of committed detachments are given a clear statement of their disciplinary authority as delegated by the higher commander.

(4) *POWs.* The handling of prisoners will depend upon the exigencies of the situation and is governed by the fact that the U.S. is firmly committed to humane treatment and care of POWs.

(5) *Burials and graves registration.* Theater army pre scribes guidance for reporting and/or marking graves within guerrilla warfare operational areas.

(6) *Morale and personnel services.* Detachment commanders' recommendations for awards are processed promptly or authority to award certain decorations is given the detachment commanders. Mail is handled by a preestablished system; automatic answers are dispatched when desired by individuals; periodic delivery may be possible with resupply drops if security considerations permit. Personal necessities are provided automatically with resupply. These normally are procured from indigenous sources or specially packaged to preserve

security and are provided for both detachment and guerrilla personnel.

(7) *Personnel procedures*. Promotion recommendations are prepared in advance to be implemented when recommended by the detachment commander. Demotion authority delegated to detachment commanders is outlined by the theater army commander.

(8) *Miscellaneous*. Policies covering pay or recognition for indigenous troops are outlined by the theater commander. When required, confidential funds are issued to the detachment commander. Barter items, such as medicine, gold, or other scarce items, are issued or held for delivery on order. Credit systems for services rendered may be established. The S1 insures that, prior to the departure of detachments for guerrilla warfare operational areas, each man is given a complete personnel processing in accordance with the SOP.

b. Distribution center. The S1 establishes and supervises a distribution center for the orderly handling of correspondence into, within and out of the headquarters. This center controls all messages except those originating. from committed detachments or TOC. The communication center operated by the communications platoon controls messages to and from committed detachments.

32. Logistics

a. General. The logistical responsibilities of the SFOB are twofold: first, support of the guerrilla warfare operational areas; second, support of the SFOB and other unconventional warfare elements as designated. To this end, the special forces group organizes a logistical support element.

The logistical support element includes organic special forces group logistics sections plus any attached or supporting logistical units from other headquarters and/or CommZ. The special forces group S4 normally controls the operations of the logistical support element. When the logistical support of the SFOB reaches such magnitude that numerous attached or supporting service elements are required, then a separate commander of the logistical support element is appointed from CommZ sources to assist the special forces group S4.

b. Organization. For a typical logistical organization in the SFOB, see figure 14.

c. Planning. The assistant S4 functions as the logistical planner for the S4 section. The assistant S4 prepares the administrative annexes for the guerrilla warfare areas operation plans. He

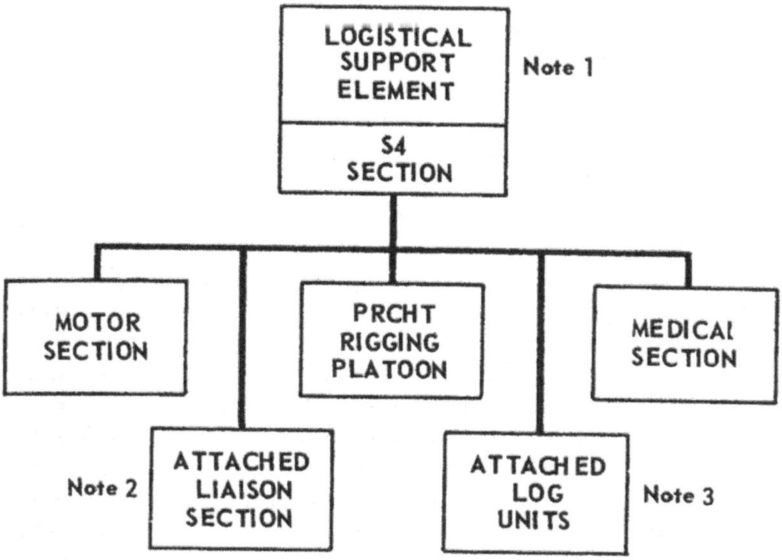

NOTES:

1. a. Receives and forwards requirements from Area Command (through Op Center) and Admin Center.
   b. Receipt, packaging, limited storage, and shipment of supplies to departure installation; or coordinates delivery, loading and movement to departure installation.
2. Logistical expediters from CommZ.
3. Logistical support units such as: Trans, Engr, Ord, Med, QM Aerial Sup, Civilian Labor.

Figure 14. SFOB logistical support element.

coordinates with S3 plans group in the preparation of these administrative annexes. The assistant S4 is the S4 representative in TOC.

d. Logistical Operations at the SFOB. The logistical support element coordinates logistical support for all elements of the base and the guerrilla warfare operational areas. The following are provided for:

(1) *Requirements.* A continuing review of requirements and submission of these requirements to the appropriate supply agency.

(2) *Limited storage.* The SFOB is prepared to provide short term storage for supplies. This storage is only temporary until supplies are packaged and shipped to the delivery

agency. Large stocks of supplies are not desirable at the SFOB ; however, nonstandard or special items are stored there for longer periods of time.

(3) *Preparation of supplies.* Normally, supplies are specially packaged for aerial delivery. Packaging is initially accomplished under the supervision of the parachute-rigging platoon assisted by uncommitted detachments. As soon as possible, QM aerial supply units assume responsibility for packaging, taking advantage of depot capabilities. Civilian labor may be employed in this operation. Packaging may expand to a sizeable operation and is accomplished in an area near the SFOB or adjacent to the departure installation. When air superiority is achieved and regular supply schedules for bulk supplies are established, the emphasis shifts from small, man-portable bundles to large packages which are heavy-dropped or air-landed. When the packaging capability exists in CommZ depots, the SFOB does not require an extensive packaging activity.

(4) *Coordination.* A liaison section attached to logistical support element coordinates with theater logistical agencies. This section, composed of CommZ and other services' logistical representatives, expedites logistical matters for the SFOB.

(5) *Support of other unconventional warfare units and attached supporting units.* The SFOB supports additional units and elements designated in plans. This may include technical service and support elements at the base plus other unconventional warfare units such as the JUWTF.

(6) *Shipment of supplies to the delivery agency.* This is normally the responsibility of the SFOB. Because of the packaging required, it usually is not feasible to ship supplies directly from depots to departure installations. If a packaging facility is located at the departure point, then supplies may be delivered direct from depot to departure installation by CommZ. If packaging can be accomplished at the depots, the flow of supplies to the departure installation bypasses the SFOB.

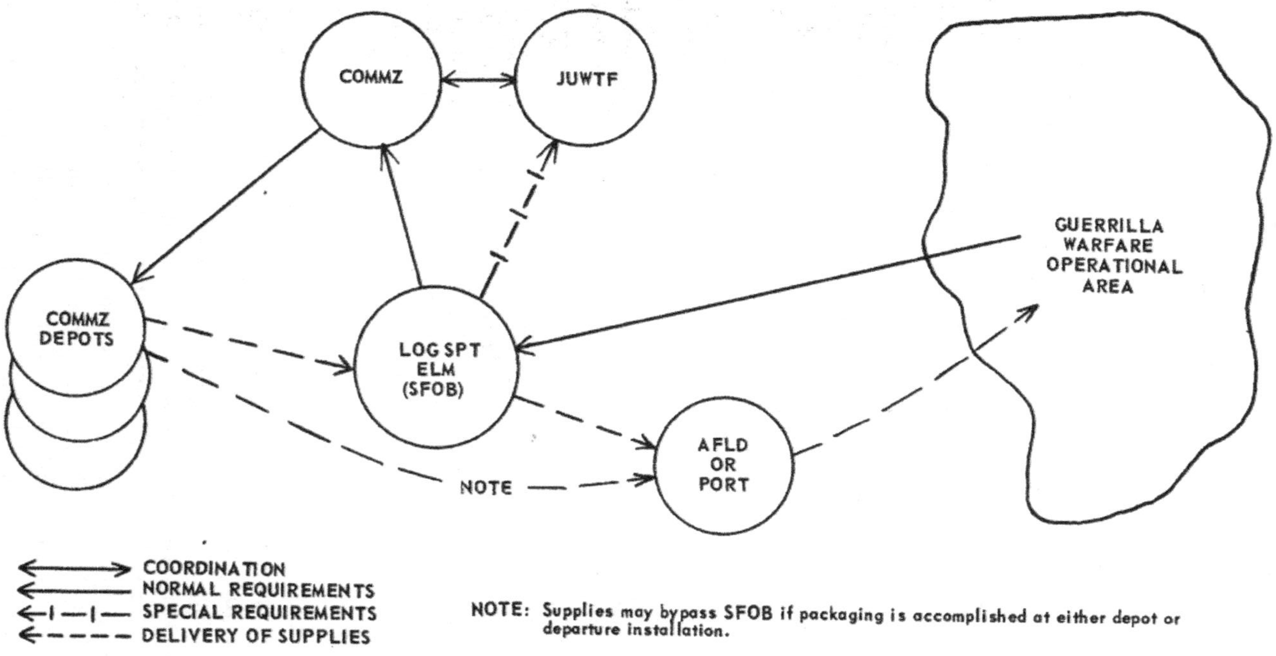

Figure 15. SFOB supply sequence.

CHAPTER 5
THEATER SUPPORT

Section I. LOGISTICS

33. General

a. Support Aspects. Within a theater of operations one or more special forces groups may be available to support the theater commander's mission. Aside from the organization precepts which were set forth in chapters 3 and 4, there are certain support aspects which are discussed from the point of view of the theater and SFOB planners. These support concepts are discussed under the general headings of logistics, intelligence, communications, and initial contact.

b. Logistics. One of the primary means used by an outside power to assist a guerrilla movement is logistical support. While guerrilla forces are expected to be able to provide a major portion of their logistical requirements from local sources, certain specialized equipment must be obtained from the sponsor. Numerous problems may be encountered in providing adequate support to guerrilla units which usually are located great distances from friendly support installations. The physical problems of transporting and delivering undetected, relatively large quantities of supplies increase in proportion to the distance into enemy-dominated territory. However, guerrilla logistical requirements are smaller in quantity and less complex than those found in a conventional force of comparable size. This tends to offset some of the disadvantages.

34. JUWTF

The JUWTF is responsible to the theater commander for the conduct of unconventional warfare, but each component service is responsible for the logistical support of its own forces assigned to JUWTF. The JUWTF prepares, or assists in preparing, the unconventional warfare annex to theater war plans, and plans and coordinates all logistical support for unconventional warfare operations in the theater. Service component commanders under the JUWTF are designated but are not in the chain of command until the JUWTF becomes operational. For this reason, JUWTF has to work directly with the logistical headquarters of each theater service component and insure the support of Army, Navy, or Air Force units that may be attached to JUWTF. The special forces group commanding officer, as commander of the SFOB, is usually the Army component commander. JUWTF headquarters provides theater army communications zone (CommZ) with logistical estimates for support of the SFOB, the committed special forces detachments and the guerrilla forces to be generated. This type of planning entails a myriad of detail and is made more complex by the many different requirements resulting from the varied geographic locations of guerrilla warfare operational areas and the varying sizes of potential guerrilla forces.

35. J4

Since JUWTF has no subordinate headquarters, troops, or logistical support capability until operations start, the JUWTF J4 usually does not supervise the physical functions of logistical support, but he is the primary staff planner and coordinator who insure that adequate means are provided. The recommendations and requests of JUWTF are incorporated into theater headquarters' plans or directives. This starts the physical process of providing the needed logistical support.

36. Theater Army

The commander of the theater army CommZ is responsible for providing administrative and logistical support to special forces units located in the theater and to other theater unconventional warfare forces as directed. The actual mechanics of procurement, receipt, storage, maintenance, distribution of supplies and equipment, and the providing of services and facilities that are needed to support the Army portion of unconventional warfare operations, differ little, if any, from the mechanics of conventional unit logistical support. Any necessary deviations are determined by direct planning and coordination between CommZ and JUWTF or the special forces group.

37. Special Forces Group

a. The special forces group or groups assigned to a theater are responsible for planning logistical support for the commitment of operational detachments and for assisting JUWTF in planning for the establishment of the SFOB. Before its attachment to JUWTF, the group is responsible for recommending that special and non-standard supplies and equipment be included in unconventional warfare plans.

b. The closest possible liaison between JUWTF and special forces group is mandatory in the following areas:

(1) Operational requirements for:

(a) Quantities of supplies and equipment.

(b) Medical support.

(c) Units and facilities for SFOB.

(2) Submission of cross-servicing requirements to JUWTF.

(3) Coordination of plans for delivery of the supplies and equipment into operational areas.

38. Logistical Doctrine and Policies

a. General. The application of logistical operations in support of unconventional warfare differs from that normally considered in support of, other theater forces.

b. Application.

(1) The five operations of logistics are—

(a) Transportation.

(b) Evacuation and hospitalization.

(c) Supply.

(d) Service.

(e) Management.

(f) Their special application to the unconventional warfare situation is described in the following paragraphs.

39. Transportation

Unconventional warfare operations can be supported by land, sea or air transportation means. As a practical matter, however, the majority of operations and situations require support by air. Because of the depth of penetration, the cargo weight to be carried and the speed with which the operation is to be executed, the Air Force has the greatest capability for this support. In certain situations, the Navy may possess a more appropriate capability for a particular operation, especially with carrier-based aircraft. This capability is utilized by the theater commander allocating the needed naval forces to provide support requested. Successful sorties with special forces detachments or cargoes require night flights of long-range and at low altitudes (50-400 feet) in order to escape electronic detection. Training of aircrews in low-level flying and navigation techniques jointly with special forces units is mandatory.

40. Evacuation

Evacuation for medical attention or other reasons from a guerrilla warfare operational area is considered only for key personnel. Even for this limited number the difficulties involved make such planning tenuous. The ability of the theater to provide transportation and the capability of guerrillas to temporarily secure evacuation sites are the governing factors.

41. Supply

a. Determination of Requirements. Proper preparation for unconventional warfare operations places a responsibility on all headquarters from special forces group level up. Initiative to commence logistical planning must be exercised at every level. A free flow of recommendations and planning guidance between all interested commands and agencies must take place. The process of determining requirements will begin when two fundamental decisions are made by theater: (1) The designation of geographical locations of guerrilla warfare operational areas, and (2) the size of the guerrilla force to be sponsored. Detailed requirements are based on the operational war plans, current logistical planning factors and miscellaneous factors such as cultural, seasonal and climatic conditions, and logistical support that may exist within guerrilla warfare operational areas. The special forces detachment commander, group staff officers, and JUWTF staff officers in. particular must be alert to determine those items of standard or, non-standard supplies or equipment needed to support operational missions. Once requirements have been determined and priorities established, requests are processed in the normal manner through army logistical channels. Local purchase procedures can hasten the procurement of non-standard or substitute items. The determination of requirements and their inclusion in appropriate war plans is a major step toward insuring adequate logistical support.

b. Stockpiling.

(1) Stockpiling is the accumulation of mobilization reserve stocks in support of strategic plans and contemplated special forces operations in guerrilla warfare operational areas. Except in unusual circumstances, items stockpiled for unconventional warfare use are stored by theater army for needs expected during the first sixty days of operations. Accessible stockpiles in adequate amounts, content, and convenient location result from thorough and detailed planning commencing with the clear determination of requirements, inclusion of these requirements in approved war plans, and the availability of funds. The creation of stockpiles for unconventional warfare operations is accomplished through the same logistical process as for other army materiel requirements. Limited stockpiling of obsolete supplies and equipment is considered appropriate for the support of unconventional warfare. Such items 'should receive minimum maintenance with the risk that only a small percentage of the materiel will not be serviceable when needed. ZI stocks of obsolete equipment should be kept on the same basis. Parts resupply will be dependent upon stockpiling.

(2) In general, unconventional warfare stockpiles are not segregated, but the items are earmarked and stored in depots along with all other similar items for theater army. Definite priorities are established by theater J4 for delivery of materiel. The problems of in-storage maintenance and inspection, as well as that of determining the best location for separate storage sites, make it a better practice to keep unconventional warfare stockpiles in CommZ depots. Logistical plans include the delivery schedule to specific locations. For exceptions to stockpiling supplies in CommZ depots, see paragraph e below.

(3) While the formation of such reserves is the responsibility of the theater commander, the content of stockpiles from the standpoints of quality and suitability must be determined by personnel in the special forces group. Small stockpiles containing only the materiel which will accompany detachments on infiltration, or be included in the initial resupply, are maintained separately and are available for emergency use.

c. Prepackaging. Prepackaging for unconventional warfare operations means that supplies and equipment destined for both initial and subsequent resupply loads are stockpiled in packages for final distribution. All of the standard supplies and equipment delivered to the operational area are packaged in one-man-portable loads of fifty pounds. Each package should contain balanced items, be complete kits for immediate use (weapon with ammunition, etc.), be safe from hazards of weather, handling, and deterioration, and have a packboard or carrying straps. Skillful use of items, such as clothing, for internal packaging material will produce savings in weight and bulk. The goal is to have packages ready for delivery. As a practical matter, however, the greater the time between packaging and actual use, the more uncertain it is that the contents will be serviceable when opened.

d. Preemergency Caches. The placement, timing, and location of preemergency caches of essential supplies to support projected operations are planned by the theater commander based upon the recommendations of the JUWTF. These caches are established when a particular need exists, adequate security can be, provided, and the packaging provides end-use serviceability of a reasonable percentage of the supplies. The many variables, such as time of use, location, security, deterioration, and the initial expense, make the establishment of each cache a matter of individual consideration.

e. Accompanying Supply Loads. Accompanying supply and initial resupply loads to 'support unconventional warfare operations are planned on a basis of austerity. The loads consist of items in quantities essential for combat operations and detachment survival for thirty days. Specific quantities of demolition materials, weapons, ammunition, and medical supplies are determined by special forces group planners from an examination of the stated mission contained in approved war plans. Once the accompanying and initial resupply equipment has been procured, it is packaged and prepared for delivery. Such materiel should not be maintained as general stock in existing CommZ depots, but located in the vicinity of departure installations.

f. Accountability. Formal accountability for supplies and equipment accompanies the materiel to the departure installation. All supplies and equipment leaving the departure installation for operational use are considered to be expended. No salvage or recovery operations are considered in logistical planning. Although all commanders concerned are responsible for the security and proper use of the materiel, the heaviest responsibility rests upon the special forces detachment commanders. SFOB will keep informal accountability for all materiel in order to report the amounts of critical items, such as weapons, committed to a guerrilla warfare operational area. Supply and distribution of critical items are controlled to assist an orderly post-war transition to peacetime pursuits.

g. Resupply Techniques. To reduce the impact of equipment losses which may occur during infiltration and subsequent operations, the SFOB schedules both automatic resupply and emergency resupply.

(1) Automatic resupply. Automatic resupply is scheduled for delivery shortly after the detachment has been committed. It is prearranged as to time, delivery site, and composition of load. This resupply is delivered automatically unless the detachment cancels or modifies the original plan. Automatic resupply is planned to replace lost or damaged items of equipment or to augment the detachment with equipment which could not be carried in on the infiltration.

(2) Emergency resupply. Emergency resupply is scheduled for delivery after the detachment has been committed and prearranged as to time and composition of load. The delivery site is selected and reported by the detachment after infiltration. The delivery of emergency resupply is contingent upon and initiated when communications from the operational detachment are interrupted for a predetermined period of time. The content of emergency resupply is normally communications and survival equipment to restore the detachment operational capability.

(3) Frequency rate of resupply. The number of resupply missions is limited until it can be determined that the detachment will not be compromised by flights over the guerrilla warfare operational area, and/or until air superiority can be established at a preselected time and place. A minimum of one resupply mission per thirty days per committed detachment is planned during this initial period. The frequency of missions increases with the degree of air superiority established by friendly forces, until resupply missions are flown as required.

(4) Catalogue supply system (app. II). In order to expedite requests, insure accuracy in identification of types and amounts of supplies and equipment, and to facilitate communications transmission security, special forces units employ a brevity code system for requesting supplies. This brevity code is known as a catalogue supply system (CSS) and its preparation is the responsibility of the C0, SFOB. The CSS is applicable to all special forces and guerrilla units. The CSS is used for three categories of supplies:

(a) Those critical items of supply essential to combat operations, e.g., arms, ammunition, and demolitions;

(b) Those critical items of supply essential for individual survival, e.g., rations, medicine and clothing;

(5) Increased amounts of the critical items contained in categories (a) and (b), but packaged in bulk for use in a rapid buildup phase of guerrilla force development.

(a) Supply bundles for categories (a) and (b) are delivered primarily during the initial phase of operations whereas category (c) bundles are delivered when the growing resistance force or tempo of combat requires increased amounts of essential combat supplies.

(b) Characteristics of an efficient catalogue supply system are critical items of supply packed in packages of fifty pounds or less; packages which are man-portable and protected from handling and weather damage; packages the composition of which is such that in storage handling and maintenance are simple; an identification code designed for accuracy in transmission when encrypted.

42. Services

Theater army emergency and war plans clearly define the responsibilities of specific CommZ service and support units in assisting the special forces group to establish the SFOB. These designated units may be attached directly to the SFOB or be assigned missions in direct support of the SFOB. Examples of service support that may be required by the SFOB are

a. Engineer (installation support).

b. Ordnance (3d echelon).

c. Medical (above dispensary level).

d. Signal (3d echelon and installation support).

e. Transportation.

f. Army aviation.

g. Counterintelligence corps.

h. Military Police (security).

i. Army Security Agency.

j. Civilian labor.

k. QM aerial supplies.

43. Management

The functions of logistics management are performed in generally the same manner at SFOB as in other military units. Management includes—

a. Supervision of the logistical operation.

b. Logistical estimates and plans.

c. Administrative annexes to orders.

d. Logistical records and reports.

e. Coordination with theater logistical agencies.

Section II. INTELLIGENCE

44. General

a. A thorough knowledge of the enemy, terrain and resistance potential, coupled with an intimate understanding of the indigenous population within operational areas, is essential to the success of unconventional warfare operations. Prior to deployment, special forces operational detachments complete detailed area studies and receive comprehensive intelligence briefings at the SFOB. After deployment, the detachment continues to add to its background knowledge by a thorough and continuing assessment of the area, using intelligence developed within the area. Thus, armed with intelligence acquired before and after infiltration, the detachment is better able to weld elements of the area command into a coordinated and effective force capable of supporting theater military operations.

b. Paragraphs 45 and 46 provide commanders with an understanding of the special forces intelligence requirements for unconventional warfare operations in general and guerrilla warfare in particular.

c. Details of procedures and techniques related to combat intelligence which are contained in the 30-series and basic branch manuals are omitted from this manual.

45. Requirements

a. General.

(1) Intelligence requirements at all command levels concerned with the conduct of unconventional warfare operations encompass the entire spectrum of intelligence. Prior to operations, the JUWTF and the special forces group both depend on strategic intelligence. As operations are initiated and special forces detachments are deployed into enemy occupied territory, combat intelligence supplements strategic intelligence. Although the area command is more often the user of combat intelligence, the SFOB and the JUWTF both use combat intelligence applicable to their level.

(2) In order to improve the chances for success in combat operations, the special forces detachment requires a greater degree of preparation in predeployment intelligence than army units of battle group or comparable size. It is desirable for the detachment to acquire this intelligence background well in advance of operations by intensive area study of predesignated regions of the world.

(3) Coordination for intelligence and counterintelligence activities in support of projected special forces operations is accomplished in peacetime by the theater army commander in accordance with joint unconventional warfare plans. During hostilities coordination for intelligence and counterintelligence activities in support of guerrilla forces is accomplished through the JUWTF.

b. Pre-Infiltration Requirements.

(1) *Operational detachments.* Special forces detachments require a thorough background knowledge of their operational area prior to deployment. This background knowledge, accomplished through the medium of aria studies, is divided into two phases

(a) *General area study.* This is the broad background knowledge of an area, region or country. See appendix III for a type general area study format.

(b) Operational Area Intelligence. This is the detailed intelligence of a designated guerrilla warfare operational area including that information necessary for the detachment to

1. Infiltrate the operational area.
2. Contact resistance elements.
3. Initiate operations. See appendix III for an operational area intelligence format.

(2) *Special forces group.* The special forces group requires current intelligence of its assigned guerrilla warfare operational areas in order to conduct preemergency planning for the employment of the operational detachments. Although area studies are prepared and provided by special research agencies, the detachment studies are reorganized into a more appropriate format for operational use. The S2 section procures the necessary intelligence documents from which detachment area studies are prepared. Coordination through prescribed channels is effected with all appropriate theater intelligence agencies for. the continuous procurement of timely area and operational intelligence. Several methods of area study preparation are feasible.

(a) The detachments prepare the general area study and as much of the operational area intelligence as is compatible with security. From the point of view of detachment knowledge, this method is the most advantageous. Disadvantages are time, security, and lack of stability in personnel assignments.

(b) The group S2 section prepares the general area study and operational area intelligence.

(c) Area specialist teams (ASTs) prepare the general area study and operational area intelligence.

(d) Various combinations of the above methods. The S2 section constantly revises area studies based upon the latest intelligence. In particular, the S2 maintains operational area intelligence as accurately and currently as possible.

(3) *Joint unconventional warfare task force.*

(a) The JUWTF requires area studies of the entire theater area of operations. Material for these area studies is provided by the theater commander. This intelligence provides the JUWTF commander with a basis for recommendations as to the selection of guerrilla warfare

operational areas that best support theater war plans. Once these areas have been approved by the theater commander, the JUWTF assigns certain areas, based upon priorities, to the special forces group.

(*b*) The JUWTF coordinates procurement of intelligence material needed by the special forces group to accomplish its operation missions. It requests from other theater component forces intelligence material and insures that requirements of the special forces group are satisfied.

c. Post-Infiltration Requirements.

(1) *Special forces operational base.*

(*a*) The SFOB is primarily concerned with intelligence which—

1. Supports the expansion of operations within active or potential operational areas.
2. Aids in determining current political trends in operational areas.
3. Aids in determining major enemy activities which influence operations within the area concerned and in other portions of the theater.
4. Aids in determining weather conditions in or en route to the area which affects external support.
5. Supports guerrilla warfare area psychological warfare operations.
6. Aids in determining enemy capabilities which could interfere with the operations of the SFOB.
7. Support cover and deception.

(*b*) The SFOB has a requirement for combat intelligence, which it reinterprets and reevaluates in light of projected theater operations.

(*c*) After deployment of special forces detachments into operational areas, the SFOB becomes an intelligence information collection agency available to all services and agencies within the theater.

(*d*) The SFOB furnishes intelligence to committed detachments as required. Much of this intelligence is provided from other theater forces.

(2) *Joint unconventional warfare task force.*

(*a*) Intelligence requirements of the JUWTF stem from planning and coordinating unconventional warfare activities within the theater. Such requirements closely resemble those of the theater commander by being broad in scope and having long-range application.

(*b*) Much of the intelligence used by the JUWTF is provided by other forces of the theater. JUWTF, through its subordinate units, is a major collection agency of strategic intelligence for the theater. It coordinates closely with the theater intelligence division and appropriate intelligence agencies of other service components for an integrated collection effort within enemy rear areas.

46. Intelligence Activities Within Guerrilla Warfare Operational Area

The deployed special forces detachment is ideally situated to contribute to the theater intelligence plan. Utilizing indigenous agencies and sources subordinate to the area command, it can gather and relay to the SFOE intelligence information of value to the theater and component force commanders. However, certain practical limitations exist as to the volume of transmission from within guerrilla warfare operational areas to higher headquarters. The security of the special forces detachment and the resistance effort restricts radio traffic and consequently the amount of intelligence information which can be expected. Higher commanders prescribe those elements of information required of special forces detachments and provide the necessary code systems to reduce the length of intelligence reports transmitted by radio. EEI assigned to special forces detachments are kept to the minimum.

Section III. COMMUNICATIONS

47. General

The communications system established for the support of unconventional warfare within a theater is designed to provide the theater commander with means to control widespread unconventional warfare forces located in denied areas. The system must also support the activities of the base elements of unconventional warfare units located in friendly territory. The JUWTF is responsible for planning and coordinating the theater unconventional warfare communications system. The special forces communications system, with its hub located at the SFOB, provides communications for special forces elements of the theater unconventional warfare effort.

48. Type Systems

a. SFOB: Theater System.

(1) The purpose of this system is to provide communications between the SFOB and other theater agencies, and between elements of the SFOB. Since the base is located in friendly territory, this communication system is comparable to the communication system of any conventional headquarters. A telephone, and teletype switchboard is provided at the base. This switchboard is connected to the area signal center and through the facilities of area signal system to all other theater and army headquarters. The special forces group has the facilities to terminate these communication lines with both telephone or teletype, either plain or encrypted.

(2) For both technical and operational reasons, the radio transmitter and receiver sites serving the base may be located at considerable distances from the base proper. Communications are provided between the operations center and the radio transmitters and receivers. The facilities of the area signal system may be used for this.

(3) Telephone service within the base proper is provided by the communications platoon of headquarters company, special forces group, but the platoon has neither the personnel nor facilities to provide long distance wire communications. The long-line wire communications are provided by the area signal center. The communications platoon does have radio and radio-teletype equipment to backup the long-line wire communications provided by the area signal system.

b. Base Command System. This system furnishes communication between the base and a detachment in an operational area. This is the communications system through which the commander coordinates and controls the guerrilla effort. It is, normally, the only link between a committed detachment and a regular military force. Since the committed detachments may be up to 2,500 miles from the base, this system must depend on radio. The radio at the SFOB can be as elaborate and as powerful as necessary. The committed detachments, however, have rigid restrictions on the size and weight of their radios. Because of this, and because of the extended distances, successful communications require more detailed planning and a higher standard of operator training than is usually the case. Since the detachment is located within a denied area, greater emphasis than normal is placed on communications security.

c. Area Internal System. This system provides the area commander with communications to subordinate elements when(1) a single special forces detachment is located in a guerrilla warfare operational area, and (2) the special forces detachment involved is a sector command subordinate to an area command. Communications in this system will initially depend on nontechnical techniques with electronic means used only in an emergency: As the area becomes more secure, the use of electronic means of communicating may be increased.

d. Area Command System. This system is established between the area command and subordinate sector commands. When the area and sector commands are separated so as to make the use of non-technical techniques impractical, radio is used. This system can be organized in any of the following ways:

(1) The sector command communicates only with the area command. The area command then communicates with the base (1, fig. 16). Emergency communication with the base is still available to the sector command.

(2) The area and sector commands have no direct communications between them. Both transmit to the base. The base relays to the field all information necessary to effect coordination (2, fig. 16).

(3) The area and sector commands have a direct communications link on operational matters. The area and sector command both communicate directly with the base on administrative matters (3, fig. 16).

e. Air-Ground System. The primary means of communication between an operational detachment and supporting aircraft is visual. Usually it is by means of a prearranged system of marking the DZ. This visual system may be supplemented, however, by electronic and/or other means.

f. Special-Purpose Systems. A special-purpose system is any system of communications established to satisfy a particular need for a definite period of time. An example of this might be the radio net established between a committed detachment and a friendly tactical unit prior to juncture.

49. Means Available

a. Within the theater of operations the communication means available to special forces are those available to other military units.

b. Between the SFOB and committed operational detachments the primary means is the radio. Other means available might include messengers, air drop of messages or a combination of messengers, and communications facilities locally available to the area of operations.

50. Factors Affecting Communications

The communications systems established to satisfy special forces requirements must be flexible. Special forces requirements can never be satisfied by a rigidly fixed system. As the operational detachments are committed and as these detachments progress in their activities, the extent and type of communications necessarily will change. Factors which must be considered are

a. The nature and extent of the resistance movement.

b. Organization of the guerrilla units.

c. The location of the guerrilla units.

d. Special forces organization of the guerrilla warfare operational area.

e. The need for communication security.

f. Secondary missions of the special forces detachment.

51. Communication Security

a. Commanders must learn and accept a concept of communications based on security and dependability, rather than speed and ease of communication. A detachment normally operates under a maximum-security type SOL This SOI does not provide for full-time two-way communications between the detachments and the FOB. The SOI provides for periodic, scheduled contacts and blind transmission broadcasts (BTBs) from the SFOB to detachments, and emergency contacts with the SFOB at any time. The enemy can be expected to make every effort to intercept and analyze detachment communications. The detachment must spend a minimum time on the air. The SOI is designed to confuse enemy intercept operators and make his intercept task as difficult as possible.

52. Radio Operator Training

The problems faced by the radio operator assigned to special forces are somewhat different than those faced by most military radio operators. A committed detachment must send messages from enemy-occupied territory. These messages travel long distances with only low-powered equipment available to the detachment. If successful communications are to be established and maintained, the radio operator must be well trained. For a soldier with no previous training as a radio operator, this training requires a long period of formal schooling. Radio operators should have the following capabilities:

a. A thorough knowledge of radio-telegraph procedures and the more commonly used operating signals.

b. A thorough knowledge of the operating procedures peculiar to special forces operations. Furthermore, he must understand how to prepare a special forces SOI and how this SOI differs from others.

c. Be capable of sending Morse Code at eighteen words per minute and receiving at twenty words per minute, using special forces operating procedures.

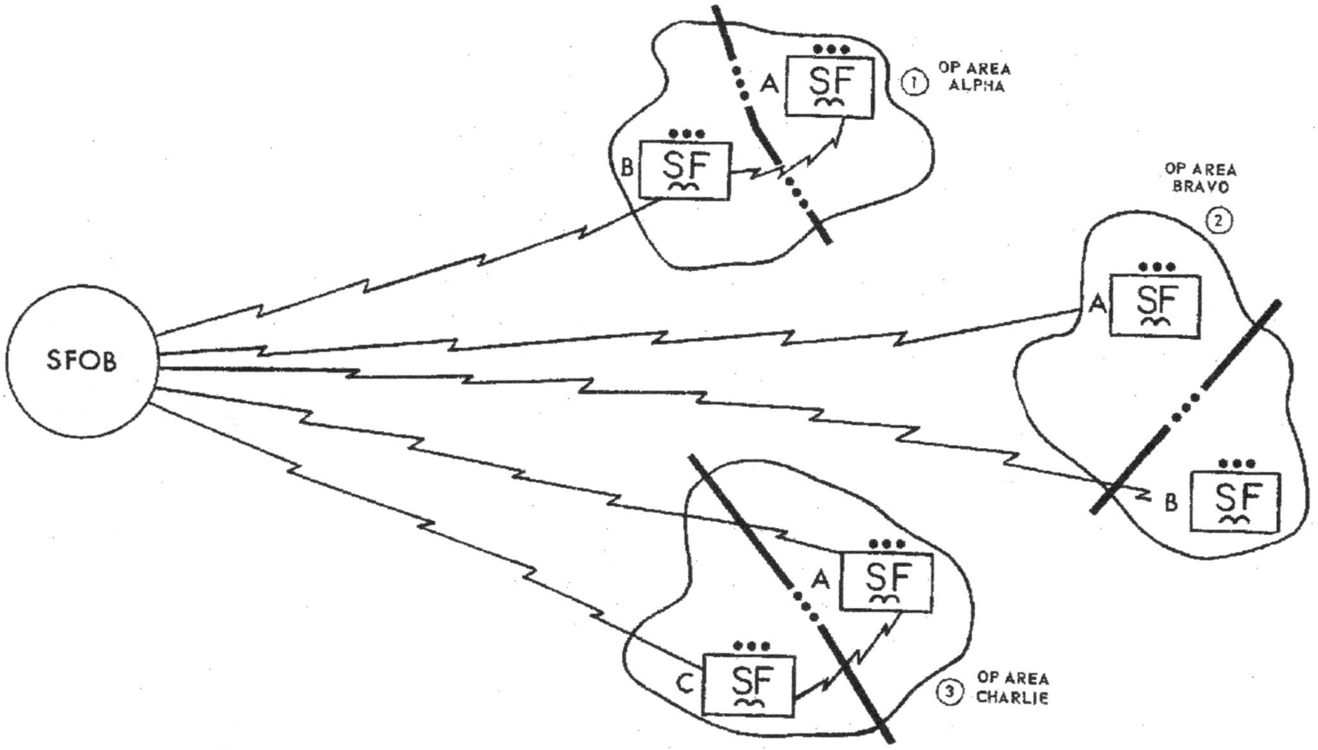

Figure 16. Area command communications.

d. Understand the principles of cryptography and be capable of devising a relatively secure system for guerrilla use. He must be proficient in the use of the primary and alternate cryptographic systems used by special forces.

e. Have a basic understanding of the theory of radio transmitters and receivers and be capable of making organizational maintenance type repairs.

f. Understand the principles of radio transmission in the HIP and UHF band.

g. Understand the construction and principles of operation of field radio antennas.

h. Understand the need for communication security. Know the principles of radio direction-finding traffic analysis and crypto analysis.

i. Have a sufficient understanding of other means of communication to be able to advise the detachment commander in their employment.

Section IV. INITIAL CONTACT

53. Contact Prior to Infiltration

a. Prior to infiltration, contact should be made with the resistance movement. It consists of the first contacts between the sponsoring power and the resistance forces. Initial contact may take place at any point in the development of the resistance force. However, to be most effective, infiltration should take place as soon as possible after the guerrilla nucleus is formed in order that special forces personnel might exert an early influence over the development of the organization.

b. The major purpose of initial contact is to arrange for infiltration of special forces detachments and for cooperation of the resistance forces with the sponsor. It permits arrangements to be made for such things as marking drop zones or coastal landing sites.

54. Contact Coincident with Infiltration

When no contact can be made in advance of infiltration but where there is evidence of guerrilla activity in an area, it may be necessary to send a detachment in "blind." In such cases, there is no reception party on the drop zone and the detachment may have only the most general idea of where to find the guerrillas. This method is the least desirable and is only used where no other alternative exists.

PART THREE
OPERATIONS

CHAPTER 6
INFILTRATION

55. General

a. The active role of special forces in the development of a guerrilla organization begins with infiltration. Infiltration is the entrance of personnel and supplies into a denied area making maximum use of deception to avoid detection. During this stage, special forces detachments leave the SFOB and are transported or moved into the guerrilla warfare operational area. Infiltration is not complete until the special forces detachment establishes communications with the SFOB from within the guerrilla warfare operational area.

b. The four means of infiltration for special forces are—

(1) Air.

(2) Water.

(3) Land.

(4) Stay-behind.

Infiltration operations may include combinations of these means.

56. Air

a. Air is usually the most practical and rapid means of infiltration. Personnel and supplies can be airlifted to virtually any place in the world.

b. The air force has the primary responsibility for providing air delivery. In most cases standard troop carrier aircraft are used. Assault type aircraft, as well as amphibious types normally associated with air-sea rescue operations, may be available and have the capability of using relatively short, unprepared airstrips. Under certain circumstances, long-range bomber aircraft is used.

c. The army can provide both rotary and fixed-wing aircraft; however, their operating ranges as well as load capacities are limited.

d. Advantages of Air.

(1) Flexibility

(2) Speed of delivery.

(3) Accuracy of delivery.

(4) Relatively short exposure to enemy action.

(5) Ability to perform concurrent missions.

e. Disadvantages of Air.

(1) Navigation precision.

(2) Vulnerability to enemy air defenses.

(3) Dependence upon favorable weather conditions.

(4) Possible injury to personnel and breakage of equipment.

(5) Possible compromise of DZ through observation of drop or electronic detection.

f. Desirable Capabilities of Aircraft. Although special forces personnel must be prepared to utilize any type of aircraft, the following characteristics and capabilities are desired

(1) Para-drop a minimum of fifteen persons and 5,000 pounds of cargo simultaneously.

(2) Operate at varying altitudes during darkness or adverse weather conditions.

(3) Possess the required navigational aids to insure locating the drop site with a minimum of difficulty.

(4) Sufficient combat radius to reach the operational area.

(5) Land and take off from unprepared airstrips with minimum useable runways of 1,000 feet.

(6) In certain situations, land and take off from water.

57. Water

a. General. Infiltration by water includes the use of surface and underwater craft. It is considered the most secure and economical means up to the point of debarkation from the parent craft.

b. Advantages of Water.

(1) Long-range of operations.

(2) Weather has little or no effect up to point of debarkation from parent craft.

(3) Evacuation is possible with each mission.

(4) Operational briefings can continue while en route.

(5) Concentration and control of personnel and supplies.

(6) Large quantities of supplies can be delivered.

c. Disadvantages of Water.

(1) Transshipment and offshore unloading are frequently required.

(2) Possibility of the loss of personnel and supplies during ship to shore movement.

(3) Movement of supplies from landing site to final destination is subject to limitations of land infiltration.

(4) Additional packaging precautions are required to protect equipment from salt water corrosion.

(5) Special training is necessary.

(6) Vulnerability to enemy shore defenses during landing operations.

58. Land

a. General. Land is generally the least desirable means of infiltration and is usually limited to short movements by individuals or small detachments. Land infiltration is best accomplished under conditions of limited visibility over difficult terrain. Land infiltration has its greatest chance of success when the enemy's lines are over-extended, the combat zone fluid, or portions of his country's borders are inadequately secured. On the other hand, a well organized, stable and closely-knit defense in depth may prohibit land infiltration.

b. Advantages of Land.

(1) Requires minimum of inter-service support.

(2) Provides an opportunity for concurrent gathering of intelligence information.

(3) Provides the ability to change routes frequently, consistent with the local situation.

c. Disadvantages of Land.

(1) Slowness with which infiltration is accomplished.

(2) Long exposure to enemy and greater , probability of capture.

(3) Limitation on amount of supplies and equipment that can be carried.

59. Stay-Behind

a. General. Stay-behind infiltration involves pre-positioning special forces operational detachments within their proposed operational areas and remaining hidden while the enemy advances through and/or occupies these areas. Stay-behind operations may be considered when the enemy has the capability of overrunning friendly areas and the attitude of the civil populace will support such operations.

b. Advantages of Stay-Behind Operations.

(1) No infiltration support required.

(2) Maximum security is obtained.

(3) Civilian contacts are already established.

(4) Caches of supplies and equipment are pre-placed.

(5) Personnel are familiar with operational area.

c. Disadvantages of Stay-Behind Operations.

(1) Proximity to enemy combat troops during their initial occupation of the operational area.

(2) Freedom of movement and communications are initially restricted.

60. Factors Influencing Choice of Means

a. Mission. The mission is the first consideration in the selection of infiltration means. A requirement for immediate interdiction will emphasize the need for speed. On the other hand, if a slower, buildup type mission is ordered, a more deliberate means of infiltration may be acceptable.

b. Enemy Situation. The enemy situation affects the means used for infiltration. For example, a heavily-guarded border may nullify land infiltration. Similarly, a strongly-defended and patrolled coastline may eliminate water as a possible choice.

c. Weather. Adverse weather conditions seriously affect air and certain phases of water operations. Conversely, bad weather may favor land infiltration.

d. Topography. Land formations must be considered in deciding on the means of infiltration to be used. Land infiltration will have a better chance for success if the chosen routes pass through mountainous or heavily-forested areas; on the other hand, these same mountains could force aircraft to fly at higher altitudes resulting in greater exposure to enemy detection and air-defense systems.

e. Hydrography. Hydrographic factors—tide data, depth of offshore water, and the location of reefs and sandbars—influence the selection of water as a means of infiltration.

f. Personnel. The number of personnel to infiltrate may be a limiting factor. The type training possessed by special forces personnel is a further consideration. In addition, if other individuals are to accompany the operational detachment, special training r-for these personnel may be necessary prior to infiltration.

g. Distance. The distance to be covered during infiltration may, eliminate consideration of certain means available

h. Equipment Available. The operational capabilities of air and water craft available for infiltration and the amount and types of special equipment—waterproofing kits, aerial delivery containers, etc.—are major factors to be considered. Limited availability in types of air or water craft will limit the number of personnel that can be infiltrated.

CHAPTER 7
ORGANIZATION AND DEVELOPMENT OF THE AREA COMMAND

Section 1. ORGANIZATIONAL CONCEPTS

61. General

The organization of a guerrilla warfare operational area involves initial organization of the area and buildup of the resistance forces. Initial organization includes establishing the required command and administrative structure, taking necessary security precautions and training a nucleus of guerrilla personnel. Buildup is the expansion of the original nucleus into an operational unit capable of accomplishing the assigned mission. Special forces detachments may infiltrate the operational area either before or after -initial organization has taken place.

62. Areas of Responsibility

a. Guerrilla Warfare Operational Area.

(1) *Designation.* The plans for organization are started when the theater commander designates certain areas within denied territory as guerrilla warfare operational areas. The guerrilla warfare operational area is the name given to a geographic area in which the army, through Special Forces, is responsible for the conduct of guerrilla warfare and other unconventional warfare activities.

(2) *Infiltration.* Any type of detachment A, B, or C may be infiltrated first. Initially, one detachment is given responsibility for the entire guerrilla warfare operational area.

b. Guerrilla Warfare Operational Sector. With the development of the area and an increase in detachments, subdivision into sectors is necessary. The sector has the same characteristics as an operational area but remains a subdivision of the area. An operational detachment becomes responsible for each sector.

63. Organization and Buildup

a. Organizational Objectives.. After infiltration, the major task is to develop resistance elements into an effective force. To facilitate this development, several tasks must be performed including:

(1) Establishment of a working command relationship between the various resistance elements and Special Forces personnel.

(2) Establishment of security, intelligence collection and communications systems.

(3) Organization of a logistical system.

(4) Provision for other administrative services.

(5) Establishment of a training program.

(6) Planning and execution of tactical operations commensurate with the state of training of guerrilla units.

(7) Expansion of forces so they are able to support theater objectives.

(8) Civilian support.

b. Area Assessment. The Special Forces detachment commander commences an area assessment immediately on reaching his area. In verifying information acquired through previous area study and briefings, he revises his plans as necessary to reflect the local situation. The area assessment serves as the commander's estimate of the situation and is the basis for plans to carry out his mission. It considers all the major factors involved, including the enemy situation and security measures, the political background of the resistance movement, and the attitude of the civilian population. There are no rigid formulae for making area assessments; each commander has to decide for himself what should be included and what conclusions may be drawn from the information he collects. The assessment is shaped by many variables including the detachment's mission, the commander's personality, and the thoroughness of the pre-infiltration study. For some detailed considerations of an area assessment, see appendix IV.

64. Command and Control

a. Organization and Control.

(1) The Special Forces detachment's primary concern is the development and control of the guerrilla forces in an area. Because the guerrilla unit is only one part of the forces generated by a resistance movement and, since the mission of Special Forces includes conduct of other unconventional warfare activities, other resistance forces—auxiliary and underground—must be considered. However, the organization and control of a guerrilla force are the essential matters for Special Forces. The other aspects of

the total resistance movement are brought in only as they bear upon the Special Forces mission.

(2) In its early stages, a guerrilla movement often is highly unorganized. Generally, the people who become guerrillas have suffered a reduction of their living standards. The main concern is grouping together for food, shelter, and mutual protection. Oftentimes several groups begin independent operations with very little concern for coordination among them. Special forces may find that guerrillas are not cooperating and may even be working at cross purposes. The Special Forces detachment must obtain control of the guerrilla groups and coordinate their actions to insure that missions assigned by the theater commander are accomplished. The degree of control varies in different parts of the world and with the specific personalities involved. As the scope of operations increases, closer coordination between guerrilla units is required.

b. Problems of Control.

(1) Although the military advantages of close cooperation between guerrilla units are obvious, a Special Forces commander may find that guerrillas resist his efforts to unify them. This opposition may be based on personal antagonisms or political or ethnic differences.

(2) A means available to the Special Forces commander to persuade guerrillas to form a united force is control of supplies. A detachment commander can use the sponsor provided supplies as a lever in convincing guerrillas to cooperate. The commander should not openly threaten to use this power, except as a last resort; but, since the guerrilla commanders are aware of its existence, it can reinforce his suggestions for unity.

(3) In his conferences with the resistance leadership, a detachment commander is careful not to become involved in their political differences. There is no surer way to lose the respect and cooperation of the guerrillas than to take sides in their internal disputes.

(4) Once guerrillas have been convinced of the advantages of close cooperation, the Special Forces commander must decide on a command structure. While he must adapt to local conditions, there are certain factors which he considers in any situation. He must have sufficient control over the guerrillas to insure that they carry out assigned missions. At the same time, the nature of guerrilla

operations requires that individual units be given a large measure of freedom in carrying out their missions.

(5) Perhaps the most delicate part of a detachment commander's job is insuring that competent leaders occupy command positions. If leaders of the original groups are not capable of filling the positions they hold, the detachment commander should arrange for their removal without creating dissension which could endanger the success of his mission.

65. Area Command

a. General. The area command is the formal organization integrating the Special Forces detachment(s) and the resistance forces within a guerrilla warfare operational area. It is established as soon as the development process requires such a step. There can be no rigid pattern for the organization of an area command. It must carry out the basic functions for which it is responsible, tailored in strength and composition to fit the situation and mission. When a guerrilla warfare operational area is subdivided, the subdivisions are called sector commands.

b. Composition. The area command basically is composed of a command group and three types of resistance forces-guerrillas, auxiliaries, and underground.

(1) *Command group.* The command group is made up of the Special Forces detachment, the local resistance leader and representatives from the resistance forces in the area. It organizes a staff as necessary. Normally, the command group is located with the guerrilla force. It is located where it can best control the resistance movement. See FM 31-21A.

(2) *Resistance Forces.* The three organizational divisions of an area command are the guerrilla force, auxiliary and underground. All three types may not be established in a guerrilla warfare operational area. The guerrilla force is the element with which Special Forces has primary concern.

c. Concept.

(1) *General.* Special forces detachments may infiltrate a guerrilla warfare operational area in different sequence to establish an area command. The order and composition of detachment infiltration depend upon many factors, some of the more important of which are—characteristics of the resistance movement, capabilities of Special Forces and needs of the theater commander. Listed below are some of

the patterns that might be followed in establishing type area commands.

(2) *Initial infiltration.*

(a) *One detachment.* One detachment-A, B, or C-may be infiltrated when the situation is not well known, the guerrilla movement is not extensive, or the guerrilla force is so well organized that minimum coordination is needed.

(b) *Multiple detachments.* Another possible solution is for two or more detachments to infiltrate concurrently, each setting up a separate sector command. This solution is adopted when topography, the enemy situation or problems peculiar to the resistance movement, prohibit the initial establishment of an effective area command (fig. 17).

(3) *Subsequent infiltration.*

(a) *Expansion from one detachment.* After an area command has been established, other detachments can be infiltrated to set up sector commands within the area (fig. 18). Either a B or C detachment is infiltrated or the initial A detachment is redesignated a B detachment. With a B or C detachment initially in the area, A detachments may be infiltrated to establish the sector commands. Subsequent infiltration of other operational detachments takes place with the expansion of the guerrilla forces, increase in operations or for political reasons.

(b) *Expansion from several detachments.* After separate commands have been established, a detachment B or C may be infiltrated to establish an area command for the same reasons as (a) above (fig. 19).

d. Control Criteria. Because of the nature of operations and the distances involved, control measures are not as effective within an area command as they are in a conventional military organization. Thus, certain criteria are established to increase effective control.

(1) *Operation order.* Sufficient guidance to subordinate units is outlined in the operation order to cover extended periods of time. This is especially true when operations preclude frequent and regular contact. Operation orders include long-term guidance on such matters as psychological operations, intelligence, target attack, air support, external

NOTE: Both operational detachments conduct operations under control of SFOB.

Figure 17. Two independent sector commands.

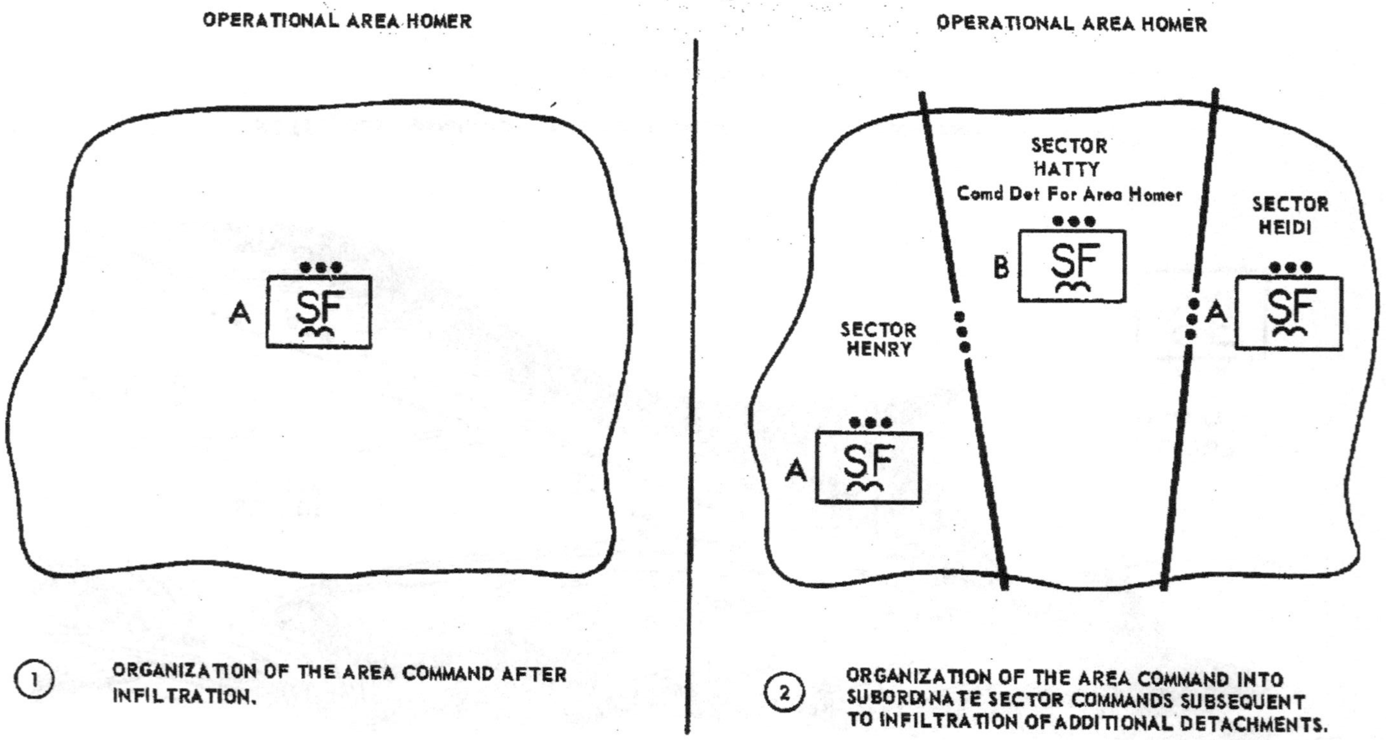

Figure 18. Guerrilla warfare operational area expanded from one detachment.

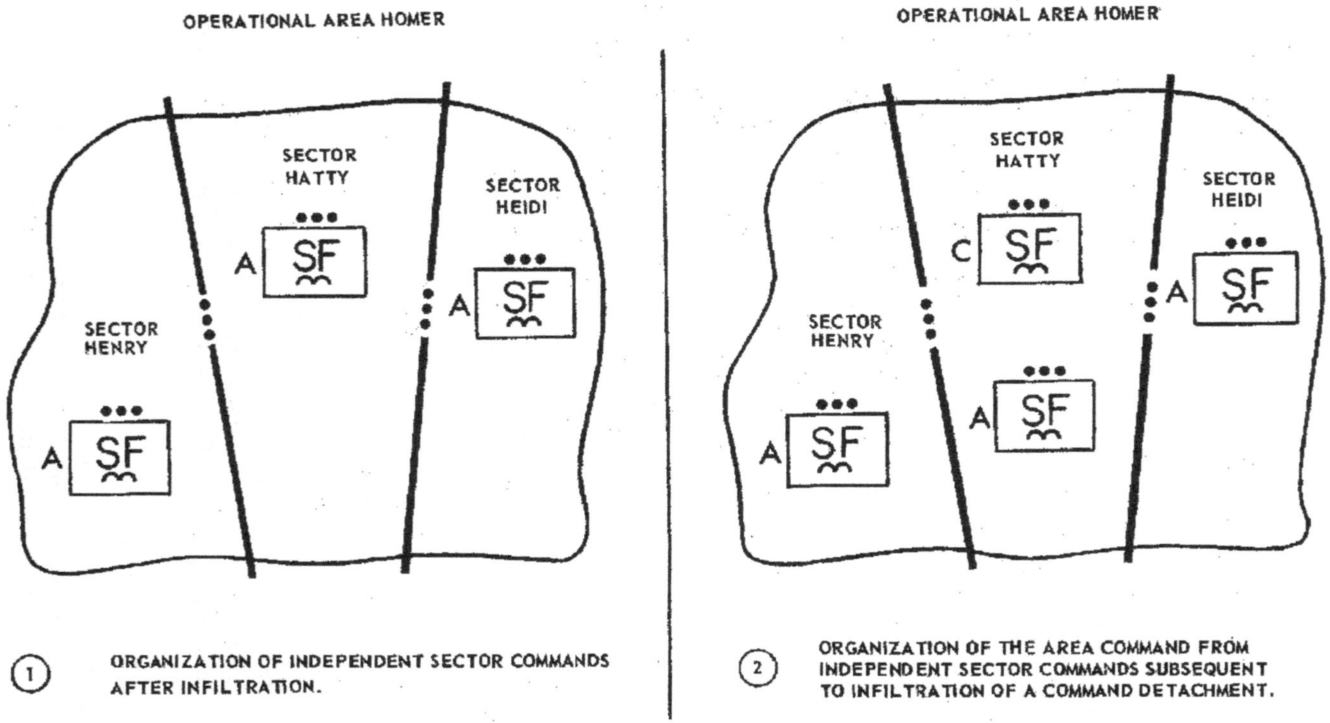

Figure 19. Guerrilla warfare operational area after initial infiltration of several detachments.

logistical support, evasion and escape, and political and military relationships vis-a-vis the resistance.

(2) *SOP's.* Another technique used to maintain control is the use of Standing Operating Procedures. SOP's standardize recurring procedures and allow the detachment and SFOB to anticipate prescribed actions when communications have been interrupted.

66. Organization on the Ground

a. General. The physical organization of the area, together with the command structure, is a priority task of the special forces commander after infiltration. In some situations the organization of the area is well established, but in others, organization is lacking or incomplete. In all cases, however, some improvement in the physical dispositions probably are necessary. Organization is dictated by a number of requirements and depends more on local conditions than upon any fixed set of rules. Among the factors considered are—degree of guerrilla unit organization; extent of cooperation among resistance forces, amount of civilian support, enemy activity, and topography. In practice, the detachment commander can expect to make compromises in organization because it is difficult to bring together in one area an ideal set of circumstances.

b. Guerrilla Base (fig. 20). The basic establishment within the guerrilla warfare operational area is the guerrilla base.

(1) *Definition.* A guerrilla base is a temporary site where installations, headquarters, and units are located. There is usually more than one guerrilla base within an area complex.

(2) *Characteristics.* From a base, lines of communication stretch out connecting other bases and various elements of the area complex. Installations normally found at a guerrilla base are—command posts, training and bivouac areas, supply caches, communications and medical facilities. In spite of the impression of permanence of the installations, a guerrilla base is considered temporary and tenant guerrilla units must be able to rapidly abandon the base when required.

c. Area Complex.

(1) Definition. An area complex consists of guerrilla bases and various supporting facilities and elements. The activities normally included in the area complex are—security and intelligence systems; communications systems; mission support sites; reception sites; supply installations; training areas; and other supporting facilities.

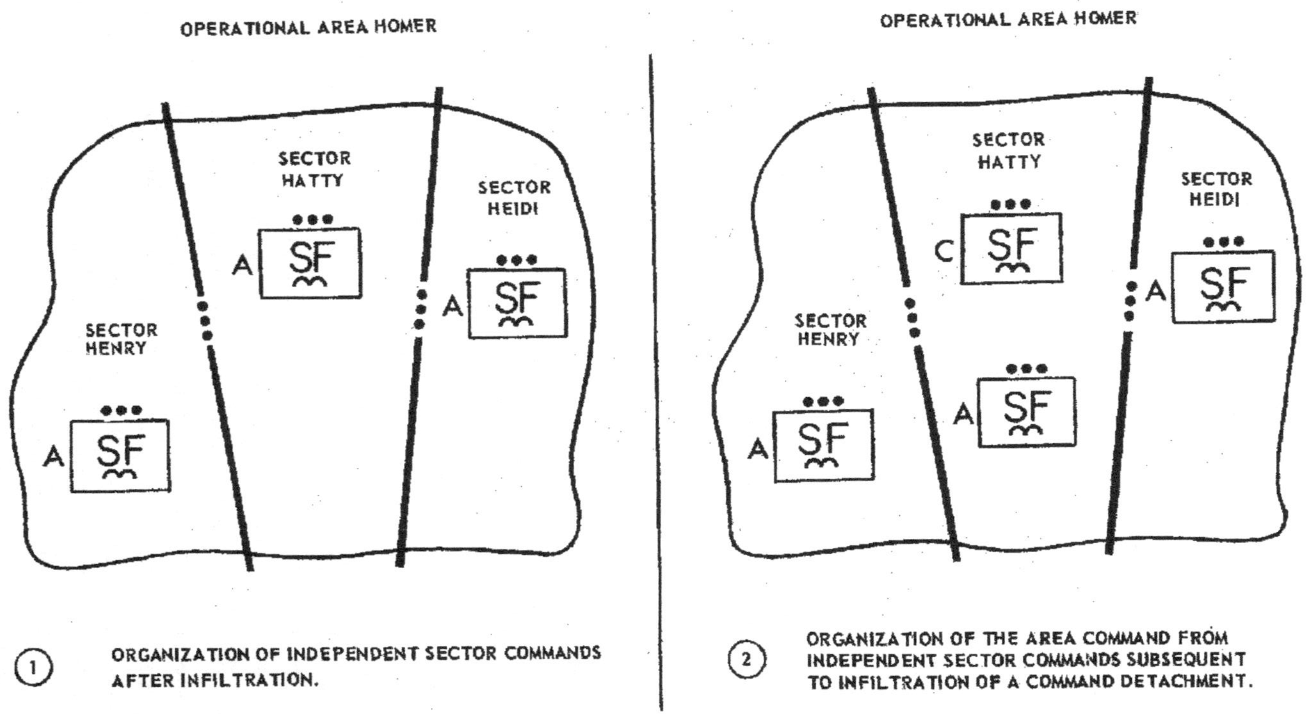

Figure 19. Guerrilla warfare operational area after initial infiltration of several detachments.

(2) *Characteristics*. The complex is not a continuous pattern of tangible installations, but may be visualized as a series of intangible lines of communications, emanating from guerrilla bases and connecting all resistance elements. The main guerrilla base is the hub of the spider web-like complex. The complex is not static but is a constantly changing apparatus within the operational area.

(3) *Location*. By virtue of their knowledge of the terrain, guerrillas should be able to recommend the best areas for locating installations. Whereas inaccessible areas are best for the physical location of guerrilla camps, the lack of these remote areas does not preclude guerrilla operations. For instance, there may be times when guerrillas are able to fight effectively in towns and on the plains. Approaches to the base are well guarded and concealed. The locations of guerrilla installations are disseminated on a need-to-know basis. Since guerrilla forces seldom defend fixed positions for extended periods of time, alternate areas are established to which the guerrillas withdraw if their primary area is threatened or occupied by the enemy.

Section II. RESISTANCE ELEMENTS

67. Guerrilla Force

a. General. Paragraph 65 refers to the three main resistance elements—guerrilla force, auxiliary, and underground that a special forces detachment will likely encounter or organize in a guerrilla. warfare operational area. The primary concern is the guerrilla force. The auxiliary and underground organizations, from the point of view of the guerrilla organization, are support elements. From the point of view of the total resistance movement, however, the guerrilla force may be supporting the under ground. For the purpose of this manual, the guerrilla force is considered the supported element.

b. Organizational Goal. The ultimate organizational goal is to integrate the guerrilla. unit and the detachment into a unified force. The degree of unification depends upon many factors. The organization which combines the special forces detachment and the guerrilla unit, regardless of the degree of cohesion, is called the area or sector command, hereafter referred to as area command.

c. Definition. The guerrilla force is the overt, militarily organized element of the area command.

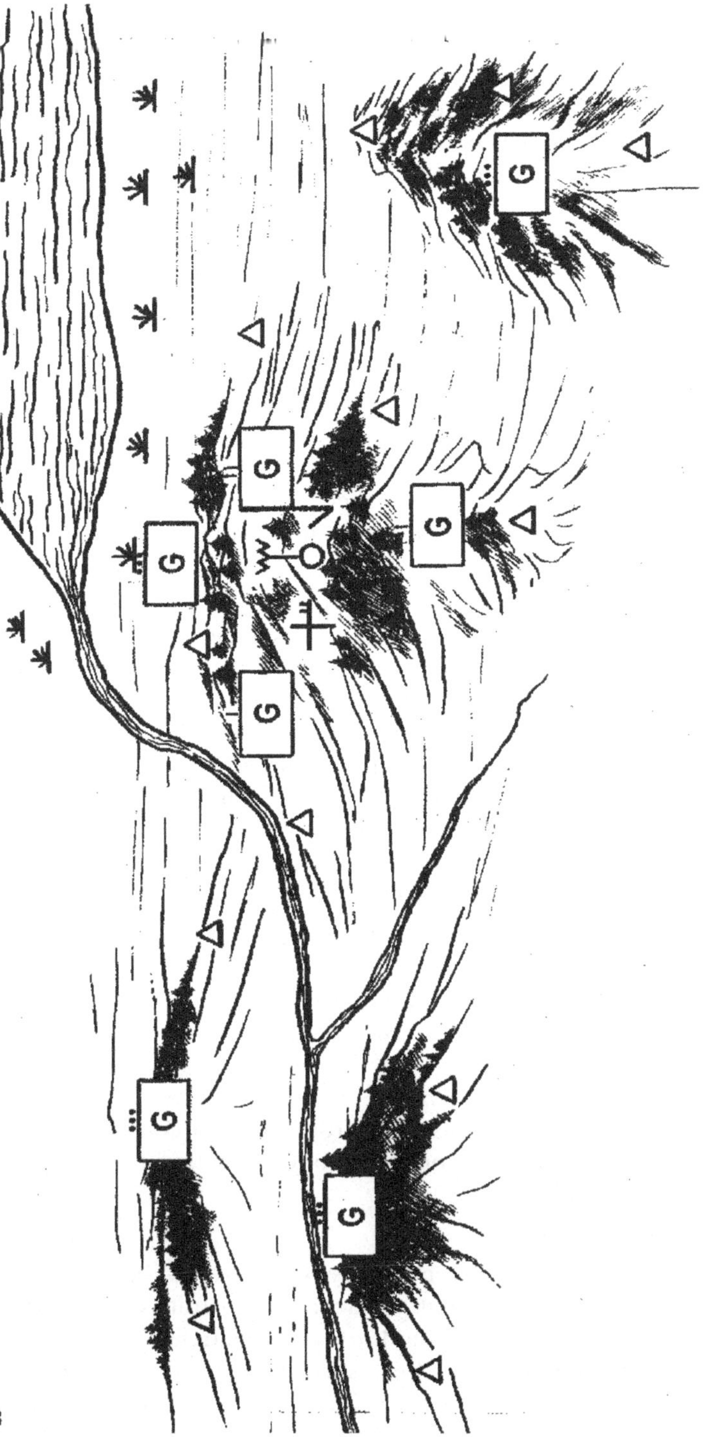

Figure 20. Guerrilla base.

d. Establishment. The guerrilla force is established when the guerrilla commander agrees to accept United States sponsorship. Once the guerrilla force is officially recognized, it is the detachment commander's responsibility to unite and control it to the best of his ability.

68. Auxiliary Fortes

(1) Active support from some of the civilian population and passive support from most of the remainder is essential to extended guerrilla operations. To insure that both active and passive support is responsive to the area command, some form of organization and control is required. Control of civilian support is accomplished primarily through the auxiliaries. Auxiliary forces compose that element of the area command established to provide for and organize civilian support of the resistance movement.

(2) "Auxiliary" is a term used to denote people engaged in a variety of activities. It is applied to those people who are not members of other resistance elements, but who knowingly and willingly support the common cause. It includes the occasional supporter as well as the hard• core leadership. Individuals or groups who furnish support, either unwittingly or against their will, are not considered auxiliaries. Auxiliaries may be organized in. groups or operate as individuals.

69. Characteristics of Auxiliaries

Auxiliary forces are characterized by location, organization and method of operation.

a. Location. Auxiliary units are composed of local civilians normally living in the smaller towns, villages, and rural areas. Unlike guerrilla units, the auxiliaries are not expected to move from place to place to conduct operations. The fact that the auxiliary forces are local and static is highly desirable from the area command viewpoint in that it provides support for the mobile guerrilla , forces throughout most of the operational area.

b. Organization.

(1) Auxiliary forces normally organize to coincide with or parallel the existing political administrative divisions of the country. This method of organization insures that each community and the surrounding countryside is the responsibility of an auxiliary unit. It is relatively simple to initiate since auxiliary commands may be established at

each administrative level, for example—regional, county, district or local (communities and villages). This organization varies from country to country depending upon the existing political structure. Organization of auxiliary units can commence at any level or at several levels simultaneously and is either centralized (fig. 21) or decentralized (fig. 22).

(2) The basic organization at each level is the command committee. This committee controls and coordinates auxiliary activities within its area of responsibility. In this respect it resembles the command group and staff of a military unit. Members of the command committee are assigned specific duties such as—supply, recruiting, transportation, communications, security, intelligence, and operations (fig. 23). At the lowest level, one individual may perform two or three of these duties.

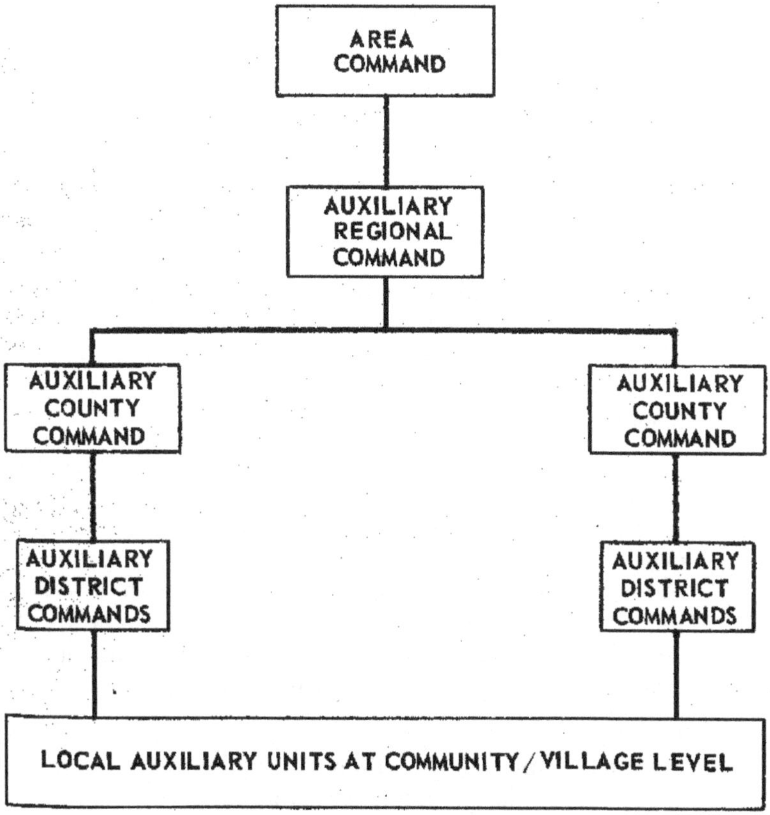

Figure 21. Centralized auxiliary organization.

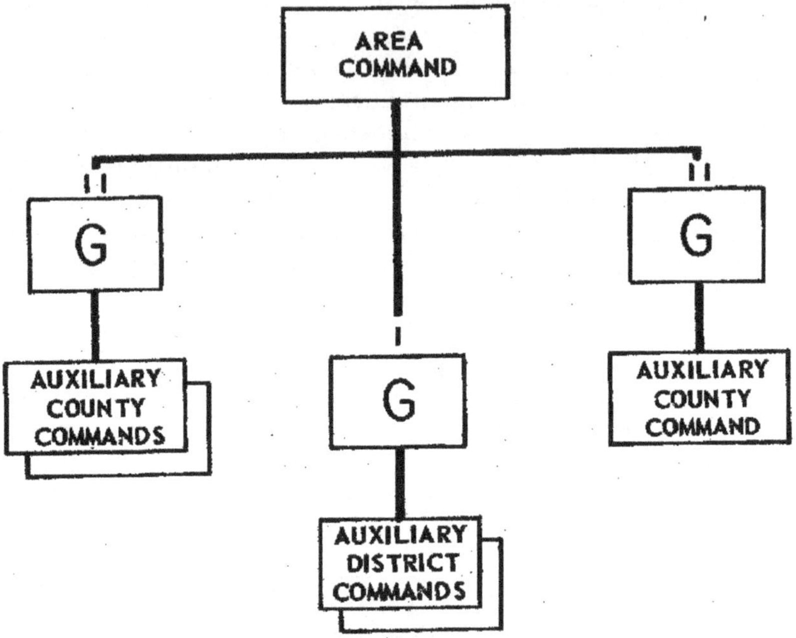

Figure 22. Decentralized auxiliary organization.

(3) The command committee may organize civilian sympathizers into subordinate elements or employ them individually. When possible, these subordinate elements are organized functionally into a compartmented structure. However, because of a shortage of loyal personnel, it is often necessary for each subordinate auxiliary element to perform several functions.

(4) The home guard is the paramilitary arm of the auxiliary force. Home guards are controlled by the various command committees. All auxiliary elements do not necessarily organize home guards. Home guards perform many missions for the local auxiliary force, such as—tactical missions, guarding of caches, and training of recruits. Their degree of organization and training depends upon the extent of effective enemy control in the area.

c. *Method of Operation.*

(1) Auxiliary units derive their protection in two principal ways—a compartmented structure and operating under cover. While enemy counter guerrilla activities often force the guerrillas to move temporarily away from given areas,

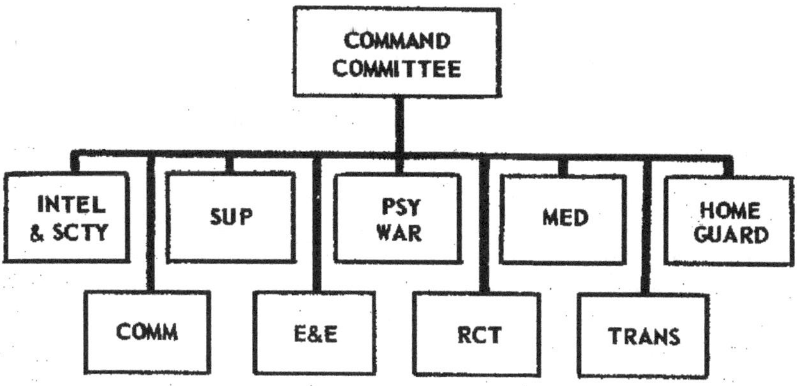

NOTE: This represents a highly organized unit separated into functional sub-units.

Figure 23. Type command committee.

the auxiliaries survive by remaining in place and conducting their activities so as to avoid detection. Individual auxiliary members carry on their normal, day-to-day routine, while secretly carrying out the many facets of , resistance action.

(2) Auxiliary units frequently utilize the passive or neutral elements of the population to provide active support to the common cause. Usually this is done on a one-time basis because of the security risks involved in repeated use of such people. The ability of auxiliary forces to manipulate large segments of the neutral population is further enhanced by the demonstrated success of friendly forces.

70. Support Missions of Auxiliaries

The support missions discussed herein are the principal ones performed by auxiliary forces to support the area command. Some of these tasks are coordinated directly with guerrilla units while others are controlled by their own higher headquarters. Normally, auxiliary units are assigned direct support missions fox guerrilla units in their areas.

a. Security and Warning. Auxiliary units provide a physical security and warning system for guerrilla forces. They organize extensive systems of civilian sympathizers who keep enemy forces under surveillance and who warn the guerrillas of enemy moves. These civilians are selected as part of the security system because of their advantageous location which permits them to monitor enemy movement towards guerrilla bases.

b. Intelligence. Auxiliary units collect information to support their own operations and those of the area command. The auxiliary force provides direct intelligence support to guerrilla units operating within their area of responsibility.

c. Counterintelligence. The auxiliary unit assists the area command counterintelligence effort by maintaining watch over transitory civilians, by screening recruits for guerrilla units, and by monitoring refugees and other non-inhabitants of the area. Because of their intimate knowledge of local people, auxiliaries should be able to report attempts by enemy agents to infiltrate the area, They can also name those inhabitants whose loyalty to the resistance might be suspect.

d. Logistics. The auxiliary unit supports guerrillas in all phases of logistical operations. They provide transportation and/or porters for the movement of supplies and equipment. Auxiliaries often care for sick and wounded guerrillas, provide medical supplies and arrange for doctors and other medical personnel. They establish and secure caches. They collect food, clothing, and other supplies for guerrilla units through a controlled system of levy, barter, ox contribution. Sometimes auxiliaries provide essential services to guerrillas such asrepair of clothing, shoes, and certain items of equipment. Auxiliary units furnish personnel to assist at drop and landing zones. They distribute supplies throughout the area. The extent of logistical support furnished by the auxiliary force depends upon the resources of the area, the degree of influence the auxiliaries exert on the population, and enemy activities. For a further discussion of logistics, see section VI.

e. Recruiting. The guerrilla units depend upon the local population for recruits to replace operational losses and to expand their forces. Auxiliaries spot, screen, and recruit personnel for active guerrilla units. If recruits are provided through reliable auxiliary elements, the enemy's chances. for placing agents in the guerrilla force are greatly reduced. In some instances, auxiliary units provide rudimentary training for guerrilla recruits.

f. Psychological Warfare. A very important mission in which auxiliary units assist is psychological warfare. The spreading of rumors, leaflets, and posters is timed with guerrilla tactical missions to deceive the enemy. Leaflets can mislead the enemy as to guerrilla intentions, capabilities, and location. The spreading of this propaganda usually involves little risk to the disseminator and is very difficult for the enemy to control.

g. Civilian Control. To control the population and give the enemy an impression of guerrilla power, the auxiliary units establish a rudimentary legal control system. This system can control black-marketing and profiteering for the benefit of the guerrilla force. Collaborators may be terrorized or eliminated by the auxiliaries. In addition, control of large numbers of refugees in the area is assumed for the guerrilla force by the auxiliary units.

h. Evasion and Escape. Auxiliary units are ideally suited for the support of evasion and escape mechanisms. Their contact with and control over segments of • the civilian population provide the area commander with a means of assisting evader.

i. Other Missions. Auxiliary units may be called upon to per= form a number of other missions to support guerrilla operation. Some of these are

(1) Activity in conjunction with the guerrillas against other targets. Cutting of telephone lines between an enemy installation and its reserve force prior to a guerilla attack is an example of such support.

(2) Operation of drop or landing zones.

(3) Operation of courier systems between widely dispersed guerrilla units.

(4) Furnishing guides to guerrilla units.

(5) Under some circumstances, they conduct active guerrilla operations in their areas of responsibility on a part-time basis.

71. The Underground

a. Enemy security measures and/or the antipathy of certain segments of the population often deny selected portions of an operational area to the guerrilla force or the auxiliaries. Since these areas are usually essential to the support of enemy operations, the resistance force attempts to extend its influence into them. The element used to conduct operations in such areas is the underground. The underground, then, is that element of the resistance force established to reach targets not vulnerable to other elements. The underground is employed to achieve objectives which would otherwise be unattainable.

b. In many respects the underground closely resembles the auxiliary force. They conduct operations in a similar manner and perform many of the same functions. The major differences, then, are twofold

(1) The underground is tailored to conduct operations in areas which are normally denied to the auxiliary force and guerrillas.

(2) The underground is not as dependent upon control or influence over the civilian population for its success. The degree to which the underground achieves this objective is a byproduct of other operations. Control of the population is not the primary underground objective.

c. For a detailed discussion of the characteristics, organization and missions of the underground, see FM 31-21A.

72. Command Group

a. The command group provides a means to control and coordinate all resistance activities in a guerrilla warfare operational area. The group normally is located with the guerrilla force. In some instances the command group may be located with the auxiliaries or the underground. For examples of the type command groups, see FM 31-21A.

Section III. SECURITY

73. General

Coincident with establishing a command organization in the guerrilla warfare operational area is the organization of an adequate security system. Security of all elements of the area command is based upon the principle of preventing the enemy from either knowing of the existence of resistance forces or being able to locate these forces when their existence is known. Security is achieved by a combination of active and passive means to include

a. The physical security warning system.

b. Dispersion.

c. Mobility of units and installations.

d. Camouflage discipline.

e. March security.

f. Communication security.

g. Counterintelligence activities.

h. Records security.

i. Cover and deception.

Since the underground and auxiliary forces achieve security by remaining undetected and through their basic cellular structure, this section deals only with the security measures applicable to the guerrilla force.

74. Bi-Zonal Security System

a. General. Guerrilla unity normally employ a bi-zonal (inner and outer zones) security system (fig. 24). There is necessarily an overlap where security responsibility devolves to two or more elements.

b. Inner Security Zone. This zone is the responsibility of guerrilla units. In this zone, the security system depends upon standard military techniques such as

(1) Patrols.

(2) Outguards and outposts.

(3) A sentinel system.

(4) Warning devices.

(5) Cover and deception.

c. Outer Security Zone. In this zone the auxiliary force and the underground provide security for the guerrillas by furnishing timely information of enemy activity.

75. Principles of Security

a. Dispersion.

(1) Guerrilla forces avoid large concentrations of troops in camps or bivouacs. Even though the logistical situation may permit sizeable troop concentrations, commands are generally organized into smaller units and dispersed. Dispersion facilitates concealment, mobility, and secrecy. Large forces may be concentrated to perform a specific operation but, upon completion of the operation, they quickly disperse.

(2) The principle of dispersion is applied to both command and support installations. A large guerrilla headquarters, for example, is divided into several echelons and deployed over a large area.

(3) In the event of well-conducted, large scale enemy operations against the guerrilla force, the area commander may divide units into even smaller groups to achieve greater dispersion and facilitate escape from encirclement. Splitting the guerrilla force into small groups is used only when all other means of evasive action are exhausted. Extreme dispersion reduces the effectiveness of the force for a considerable period of time. This action also lowers the morale of the guerrillas, and weakens the will of the

SENTINELS PATROLS OUTPOSTS AND OUTGUARDS

Inner Zone

Figure 24. Bi-Zonal security system.

Ⓐ

Ⓑ

Ⓒ

Ⓔ

AUXILIARIES MAINTAIN CONTINUOUS SURVEILLANCE OF THE ENEMY Ⓐ Ⓑ Ⓒ AND REPORT Ⓓ HIS ACTIONS AND MOVEMENTS TO THE GUERRILLAS Ⓔ.

Ⓓ

Outer Zone

Figure 24—Continued.

civilians to resist. To increase the probability of successful reassembly of dispersed units, plans include alternate assembly areas.

b. Mobility.

(1) Guerrilla installations and forces maintain a high degree of mobility. Evacuation plans for installations and forces include elimination of all traces of guerrilla activity prior to abandonment of the area.

(2) Mobility for evacuation is achieved by preparing equipment to be moved in one-man loads, by caching less mobile equipment, by destroying or hiding material of intelligence value to the enemy, by policing the area, and by eliminating signs of the route of withdrawal.

c. Cover and Deception Operations. Another principle of providing security is the use of deception operations. These operations are planned to deceive the enemy as to location, intent or strength of the guerrilla force. They are conducted in conjunction with other resistance operations in the area or with conventional force operations.

76. Security Discipline

a. Safeguarding Plans and Records.

(1) Information concerning guerrilla operations is disseminated on a need-to-know basis. Minimum necessary copies of documents are made or maintained. Each person is given only that information he needs to accomplish his tasks. Special efforts are made to restrict the amount of information given to individuals who are exposed to capture.

(2) Administrative records are kept to a minimum and cached so that the location is known only by a required few. Usually essential records are photographed.

(3) Whenever possible, reference to names and places are coded and the key to the code is given on a need-to-know basis.

(4) Records which are of no further value are destroyed.

(5) The guerrilla relies on his, memory to a far greater extent than the regular soldier. Installations are not marked on maps or papers which are taken out of the base. Guerrillas habitually memorize the location of installations and areas to which they have access.

b. Security Measures.

(1) Strict security measures are enforced. These include

(a) Camouflage discipline.

(b) Isolation of units from each other.

(c) Proper selection and rigid supervision of courier routes between headquarters and units.

(d) Police of camp sites and installations.

(e) Movement control within and between guerrilla bases.

(f) Isolation of guerrilla. units from the civilian population at large. Any necessary contact with civilians is accomplished through auxiliary elements.

(g) Thorough indoctrination of all units in resistance to interrogation.

(2) Security consciousness is impressed upon guerrilla. troops from the inception of the force and continues throughout operations. Commanders at all levels constantly strive to improve security measures. Particular attention is devoted to those units and elements that have recently been inactive or are located in relatively safe areas.

77. March Security

a. Security on the march is based upon accurate knowledge of he enemy's location and strength. The intelligence section of the area command provides this vital information for security of movement.

b. Once routes have been selected, units are briefed on enemy activity, primary and alternate routes, dispersal and reassembly areas along the way, and security measures to be observed en route. Auxiliary units in the route area assist by providing security elements for the guerrillas.

c. While on the move, the guerrilla forces employ march security techniques such as advance, rear and flank guards. Pre-selected bivouacs are thoroughly screened by patrols prior to their occupation by guerrilla units. Contact is established with local auxiliary units designated to support the guerrilla movement. The auxiliaries are thus able to furnish the latest enemy information to guerrilla commanders.

78. Counterintelligence

a. Security measures used by guerrillas to safeguard information, installations and communications, are supplemented by an active counterintelligence program to neutralize the enemy's intelligence system and to prevent the penetration of guerrilla forces by enemy agents.

b. Counterintelligence is a command responsibility under the staff supervision of the intelligence section of the area command. Selected personnel, specially trained in counterintelligence, carefully screen all members of the guerrilla organization as a protective measure against enemy infiltration. They also plan and supervise an active campaign of deception.

(1) Counterintelligence personnel through the auxiliary forces keep a constant check on the civilian population of the area to insure against the presence of enemy agents within their midst. Civilians upon whom the guerrillas depend heavily for support may compromise the guerrilla warfare effort as easily as a disloyal guerrilla.

(2) False rumors and false information concerning guerrilla strength, location, operations, training and equipment can be disseminated by counterintelligence personnel. Facts are distorted intentionally to minimize or exaggerate guerrilla capabilities at any given time.

(3) Active measures are taken to determine enemy intentions, methods of operation, EEI, and to identify enemy intelligence personnel or local inhabitants who may be used as enemy agents. These active measures include penetration of enemy intelligence and counterintelligence organizations by selected personnel, and the manipulation of defectors and double agents.

79. Security Role of the Auxiliary and Underground

Both the auxiliary forces and the underground contribute to the security of the guerrilla force. Incidental to their everyday operations, they uncover enemy activity or indications which, when evaluated, disclose potential danger to the guerrilla force. They establish specific systems designed to provide warning of the approach of enemy units. They intimidate any collaborators and attempt to elicit information from enemy personnel, local officials and the police. They operate in what is to the guerrillas the outer security zone.

80. Reaction to Enemy Operations

Premature or unnecessary movement caused by the presence of the enemy may expose guerrillas to greater risks than remaining concealed. Such moves disrupt operations and tend to reduce security by exposing guerrillas to enemy agents, informants, and collaborators. The decision by the guerrilla commander to move is made only after a careful estimate of the situation.

Section IV. INTELLIGENCE IN GUERRILLA WARFARE OPERATIONAL AREAS

81. General

The location of the area command in enemy-controlled territory makes available to the theater commander an additional means of developing intelligence. The area command is able to exploit sources generally unavailable to other theater forces. However, the area command is not primarily an intelligence agency but a military force responsive to theater control. As such, they provide intelligence information gathered incidental to their .primary mission.

82. Agencies and Sources

a. The area command has available three principal agencies to assist in gathering intelligence information. They are the guerrilla force, the auxiliary force and the underground.

b. For special intelligence collection missions, the special forces detachment may be augmented by trained intelligence specialists.

83. Capabilities

The area command has the following intelligence capabilities:

a. Intelligence to Support Their Own Operations. The intelligence system of the operational area is primarily geared to support the command. It produces intelligence for the use of the guerrillas, the underground, and the auxiliaries.

b. Intelligence Data to Support Theater Operations. The area command in the course of operations acquires intelligence data of value to other theater forces. Some examples are

(1) Order-of-battle intelligence data.

(2) Information to support psychological warfare activities.

(3) Target information for tactical and strategic air forces plus post-strike information.

(4) Information of political, sociological, and economic intelligence value.

(5) Intelligence data to support specific tactical operations such as airborne, amphibious, or armored operations.

Section V. COMMUNICATIONS IN GUERRILLA WARFARE OPERATIONAL AREAS

84. General

Communications within an area or sector furnish the commander the means to control his organization. Because it is located in enemy territory, the communication system will be slower. When a plan is formulated, more time must be allowed for transmitting orders than in conventional military units.

85. Means Available

Communications within a sector or between the area and sector commands use nonelectronic techniques wherever practical. Until the area is relatively secure, electronic means should be used only when absolutely necessary. Since the area command is an integrated organization consisting of the special forces detachment and the resistance forces, the same communications must satisfy the requirements of both. Special forces communications at the detachment level cannot be divorced from guerrilla communications.

86. Communication Security

a. Communications are always vulnerable to interception. Absolute security does not exist. Interception of communications is an excellent means of gathering intelligence information. Special forces radio operators normally operate with a maximum-security SOI, designed to make interception of electronic communications difficult. If guerrilla radios are operated from the same general area as the special forces radio, they operate using the same precautions.

b. During the early phases of guerrilla development, messengers are the chief means of communication. Security is enhanced by cellular structure of the messenger organization, use of secure cryptographic systems and proper authentication.

87. Factors Affecting Communications

Radio used between the area and sector commands is the radio operated by trained special forces operators. Communications within an area or a sector depend on the operational situation, the physical location of the area and sector commands, terrain barriers, the training of the resistance force, the enemy capability in electronic interception, the security of the area and the communication equipment available. The range of radios, which operate in the high frequency band, is extremely difficult to predict. Under certain conditions these transmissions can be intercepted over great distances. The range of low-powered radios, operating in the VHF band, rarely exceeds line of sight. Until the area is secure, the use of radios is restricted to those operational missions from which little intelligence data will accrue to enemy interceptors. Enemy capability to intercept either electronic or nonelectronic communications and the operational situation are the two primary factors to be considered when planning communications within an operational area.

Section VI. LOGISTICS IN GUERRILLA WARFARE OPERATIONAL AREAS

88. General

a. The logistical support for guerrilla forces is derived from two primary sources; the sponsor and the operational area. Logistical planning of the area command is based upon resources available from both of these sources.

b. The operational area is normally expected to provide the bulk of the logistical support required by the area command. This support includes local transportation, care of the sick and wounded, various services, and those items of supply necessary for day-to-day existence such as food, clothing, and shelter. During the course of operations, the area may be able to provide a certain portion of arms and equipment. This materiel is usually procured as a result of combat action against the enemy or security and police forces. In highly developed areas of the world, certain amounts of technical equipment, such as radios, can be locally obtained.

c. The sponsor provides those essential logistical services which are not readily available within the operational area. Usually this consists of arms, ammunition, demolitions, and communications equipment—the essentials to support combat operations. Under certain conditions, sponsor logistical support is expanded. It then includes evacuation of the sick and wounded, food, clothing, and other essential survival items unavailable in the area.

89. Logistical Requirements

Logistical requirements of the area command are rudimentary and simple when compared to a conventional force of similar size. These requirements, in general, consist of—

a. Necessities to enable guerrillas to live; such as food, clothing and shoes, shelter, and medical equipment.

b. Combat equipment for the conduct of operations. Major categories are—arms and ammunition, demolitions and communications equipment.

c. Sufficient transportation to enable guerrilla units to distribute supplies.

d. A medical system to care for sick and wounded.

e. Essential services, for example—the repair of shoes and clothing.

90. Logistical Organization

a. Area Command. The area command organizes for logistical support by assigning tasks to its subordinate elements.

b. The Guerrilla Force. Each guerrilla unit is assigned a portion of the operational area for logistical support. Usually guerrilla units are satellited on an auxiliary region and receive direct logistical support from the auxiliary units within their assigned portion of the operational area. In addition to the support from local auxiliaries, the guerrilla unit depends upon its own avert action to satisfy logistical requirements.

c. The Auxiliary Force. One of the primary roles assigned to auxiliary units is logistical support of guerrilla units. Since the auxiliaries themselves are largely self-sufficient because they live at home, they establish local logistical systems designed to support guerrilla units.

d. The Underground.

(1) The underground logistical role is largely one of self-maintenance for its own members. This usually takes the form of ration cards, documents, money, living quarters and special equipment. In the latter case, the area command often supplies special equipment received from the sponsor.

(2) In some situations, the underground provides selected items of supply, which would otherwise be unobtainable, to the area command. Examples are—drugs and other medicines, radios, raw materials for explosives, photographic materials, etc.

91. Supply

a. External Supply.

(1) Supply of the area command from external sources is normally limited to those items not readily obtainable in the operational area. Depending on conditions within the area, this varies from small, irregular deliveries to total logistical support.

(2) As a general rule, sponsor-provided supplies are delivered directly to the individual user. For instance, if two guerrilla battalions are separated by a distance of twenty miles, the supplies for each are air-dropped on separate drop zones selected to service each battalion.

(3) The situation may be such that direct delivery to the user is not desirable or possible. In this case, supplies are delivered to a designated location and their contents distributed to the various users. Although this system takes much time and effort, it permits centralized control over sponsor-provided supplies and is the preferred method when the situation requires close supervision of subordinate elements.

b. Internal Supply. This system includes all the methods used by the area command to obtain supplies and equipment from within the operational area. In dealing with the civilian population, the resistance elements must balance their requirement for supplies against the need to maintain cooperation of their civilian supporters. A procurement program designed without regard to the needs of the population may impose such heavy commitments on the civilians that they refuse to cooperate and thus limit the operations of the resistance force and increase the requirement for external supply.

(1) *Offensive Operations and Battlefield Recovery.* By conducting offensive operations against the enemy, the guerrilla force is able to satisfy many of its logistical requirements and at the same time deny the use of these supplies to the enemy. Capturing supplies from the enemy has the advantage of not alienating civilians. With adequate intelligence and proper planning, raids and ambushes are conducted against installations and convoys containing the items needed by the guerrilla force. Prior to an operation, each guerrilla is instructed to secure those priority materials required by the guerrilla force. In areas where conventional operations have been conducted, guerrillas can obtain certain quantities of supplies by collecting abandoned equipment.

(2) *Levy.* To ensure an equitable system for obtaining supplies from the local population, a levy system based on the ability of each family or group of families to contribute may be organized. This system is established and operated by the various auxiliary units. Such a system provides a means whereby the burden of supplying the guerrilla force is distributed equitably throughout the civilian population. The population can be told that payment will eventually be made for the supplies taken. Receipts may be given to individuals or records of the transactions kept by the area command supply officer. In establishing the levy system,

the commander must consider a number of obstacles which might affect procurement in his operational area. Among them—

(a) Chronic food shortages.

(b) Enemy interference and/or competition for supplies.

(c) Possible impact of combat actions, such as "scorched earth" policies and radioactive contamination.

(d) Competition from rival guerrilla bands.

(3) *Barter*. It may not be desirable for the area command to engage in outright barter with the civilian population because of possible adverse effects upon the levy system. However, in some cases it is mutually beneficial to exchange critical items, such as medical supplies, for food, clothing or services.

(4) *Purchase*. Special forces detachments may be given a certain amount of negotiable currency in the form of gold or paper money. This money is for the purchase of critical items or services within the operational area. There will not be enough for purchases to meet all supply requirements. In addition, the uncontrolled injection of large amounts of money may well disrupt the local economy. Purchases are used to supplement rather than replace the levy system.

(5) *Confiscation*. Confiscation is a method which may be used to fulfill those requirements which cannot be met by the other methods of internal supply. Confiscation is often employed in cases where certain groups refuse to cooperate or are actively collaborating with the enemy. Naturally, confiscation tends to alienate the civilian population and therefore should be used only in emergencies or to punish collaborators. In all cases, confiscation must be strictly controlled to insure that it does not deteriorate into indiscriminate looting.

c. *Storage*

(1) The storage or caching of supplies and equipment plays an important role in the area command logistical plan. The uncertainties of the weather and enemy action prevent timetable receipt of supplies from the sponsoring power. The area command, therefore, must be prepared to operate for extended periods without external resupply. This necessitates stockpiling supplies for later use. Guerrilla

units do not maintain excess stocks of supplies since large quantities of equipment limit mobility without increasing combat effectiveness. Supplies in excess of current requirements are cached in a number of isolated locations to minimize the risk of discovery by the enemy. These caches are established and secured by both guerrilla and auxiliary units in support of the guerrilla force. Items are carefully packaged so that damage from weather and exposure is minimized. Specialized packaging of supplies is accomplished by the sponsor.

(2) Caches may be located anywhere that material can be hidden-caves, swamps, forests, cemeteries and lakes. The cache should be readily accessible to the user. Dispersal of caches throughout the operational area permits a high degree of operational flexibility for the guerrilla force.

(3) Generally there are two types of caches those containing items used on a day-to-day basis and those containing items to be used in the future. Each unit caches excess supplies and equipment and draws upon these as needed. Only the unit commander and key personnel know the location of caches. In the same fashion, commanders establish caches containing supplies which represent a reserve for emergency use throughout the area.

92. Transportation

a. The transportation requirements of the area command are met largely from within the area since it is usually impractical for the sponsor to provide transportation support for operational use. To fulfill its transportation requirements, the area command utilizes any means available.

b. Movement by foot is usually the primary means, especially in the initial stages of guerrilla development. In specific situations, this may be supplemented by locally-procured motor vehicles or animals. The auxiliaries provide whatever local transportation is available to guerrilla units. This transportation is normally furnished on a mission basis. However, in some instances, the guerrillas permanently acquire transportation and organize supply trains.

93. Medical Service

a. Area Command Medical Requirements. The area command medical requirements vary widely between operational areas but usually differ from the conventional medical problem in two respects. First, due to the nature of guerrilla operations, battle casualties are normally lower in guerrilla units than in their infantry counterparts.

Second, the incidence of disease and sickness is often higher in guerrilla forces than in comparable conventional units.

b. Area Medical Support System. The area medical support system is based primarily upon local facilities supplemented by sponsor-provided medical supplies.

c. Medical System in the Operational Area. The medical system in the operational area features both organized guerrilla medical units and auxiliary medical facilities for individuals and small groups. The former are located in guerrilla base areas and staffed by guerrilla medical detachments. The auxiliary facility is a location in which one or a small number of patients are held in a convalescent status.

d. Guerrilla Medical Detachment.

(1) Regardless of the varying size of guerrilla units, the medical detachments retain essentially the same structure and functions. Their duties are to maintain a high state of health in the command, to render efficient treatment and evacuation of casualties, and to insure the earliest possible return to duty of those who are sick or injured. The detachment may also provide treatment and drugs to auxiliary and underground elements.

(2) The organization of the medical detachment consists essentially of three sections—the aid station, which is charged with the immediate care and evacuation of casualties; the hospital, which performs defensive treatments of casualties and coordinates medical resupply and training; and lastly, the convalescent section, which cares for patients who require rest and a minimum of active medical attention before their return to duty. The convalescent section is not located near the hospital area as this increases the size of the installation and thus the security risk. Instead, the patients are placed in homes of local sympathizers or in isolated convalescent camps.

(3) During the early stages of development, the medical organization is small and probably combines the aid station and the hospital into one installation. The use of auxiliary convalescent facilities is found at all stages of development.

e. Evacuation.

(1) Every effort is made to evacuate wounded personnel from the scene of action. The condition of wounded guerrillas may preclude movement with the unit to the base. In this event, the wounded are hidden in a covered location and the

local auxiliary unit notified the local auxiliaries then care for and hide the wounded until they can be returned to their own organizations.

(2) The evacuation of dead from the scene of action is most important for security reasons. The identification of the dead by the enemy may jeopardize the safety of their families as well as that of their units. The bodies of those killed in action are evacuated, cached until they can be recovered for proper burial, or disposed of by whatever means is consistent with the customs of the local population.

(3) As the operational area develops and the overall situation favors the sponsor, evacuation of the sick and wounded to friendly areas may be feasible. This lightens the burden upon the meager facilities available to the area command and provides a higher standard of medical care for the patient.

f. *Expansion of Medical Support.*

(1) As the arms, command expands, ft is more efficient from a medical standpoint to establish a centralized system to provide advanced medical care. Field hospitals permit more flexibility because of them wider selection of trained personnel, equipment to provide special treatment, and they relieve the aid stations oaf the responsibility for prolonged treatment of patients. Since this type of installation may be fairly large and may have sizeable amounts of equipment, its mobility will suffer. For that reason it is located in a relatively isolated area away from troop units, headquarters and other sensitive areas but so as to receive the maximum protection from guerrilla units.

(2) To prevent the hospital from becoming so large that it attracts undue attention, certain actions are taken. First, as trained personnel, supplies and equipment become available, additional hospitals are established. Second, as soon as possible, a patient is transferred to a convalescent home to complete his recovery, If the individual is placed in a civilian home, he is properly documented.

(3) In some cases the local population may not be able to support the area command With qualified medical personnel. As the requirement for doctors and specialized personnel increases, the SFOB may have to provide additional medical personnel over and above the detachment's organic medical personnel.

94. Services

In guerrilla warfare operational areas, services are primarily restricted to basic maintenance and repair of equipment. The difficulties in procuring supplies dictate the need for rigid supply discipline. All personnel must perform first-echelon maintenance. Plans provide for the maximum utilization of available supplies and the establishment of local repair facilities to prolong the life of equipment. Necessary maintenance and repair items such as armorers tools, small arms repair kits, sewing kits, oil and cleaning materials are included in sponsor-provided supply packages. Clothing and footgear are repaired locally.

CHAPTER 8
COMBAT EMPLOYMENT

Section I. INTRODUCTION

95. General

a. Although discussed separately from other operations, the combat employment of guerrilla forces commences early and continues throughout the entire span of guerrilla warfare development. However, combat employment normally reaches its peak just prior to. the juncture between unconventional and conventional forces.

b. Control and coordination of guerrilla units. is assisted by the designation of guerrilla warfare operational areas. The subdivisional concept of these areas is explained in paragraph 62.

c. Guerrilla forces have a much greater chance for success and most effectively support conventional military operations when their activities are coordinated with other theater forces. Coordination of the guerrilla effort with the service component commands' plans of operations is executed through normal command channels-theater commander, JUWTF, SFOB, and operational detachment.

96. Area Control

Guerrilla forces are rarely concerned with seizing and holding terrain. However, they are concerned with establishing area control in order to expedite operations. Area control is classified, according to degree, as area superiority or area supremacy.

a. Area Superiority. Temporary control of a specific area is attained through maximum use of the principles of surprise, mass, and maneuver. Area superiority is maintained only for the period of time required to accomplish missions without prohibitive interference by the enemy.

b. Area Supremacy. Complete area control is attained whenever the enemy is incapable of effective interference with guerrilla operations. Area supremacy is seldom achieved through unconventional warfare efforts alone.

97. Nature of Guerrilla Warfare

a. No word describes the nature of guerrilla warfare better than "fluid." In guerrilla warfare the situation is always fluid. Both enemy and guerrilla units move and change their relative positions as the result of tactical maneuvers. The area of guerrilla activity is never static; the situation changes constantly as the enemy reacts to guerrilla actions.

b. Maximum effective results are attained through offensive operations of the guerrilla force. Normally, the guerrilla force is primarily interested in the interdiction of lines of communications and destruction of critical enemy installations. Except in those instances wherein the tactical advantages are clearly with the guerrilla force, no effort is made to close with and destroy ail enemy. Conversely, the enemy force must provide security for his critical installations and seek to contact and destroy the guerrilla force. These opposing courses of action create ail operational environment that is fluid.

c. Guerrilla area superiority is more easily achieved in difficult terrain that restricts enemy observation and movement. These factors reduce the enemy capability to mount coordinated operations quickly against the guerrillas and allows sufficient time for guerrilla units to avoid becoming involved in static defensive combat. The physical characteristics of these so-called "redoubt areas" are usually such that critical enemy targets are not located in areas of sustained guerrilla superiority.

d. The enemy can achieve area superiority or supremacy of a particular region at any time he is willing and able to commit sufficient forces to do so. However, because the guerrilla force is comparatively free to select the time and place of attack, successful operations are conducted against target systems despite enemy security measures.

e. In between those areas of enemy control and temporary guerrilla force control is an area or twilight zone subject to permanent control of neither. Because the area command call initiate offensive operations employing a variety of methods of attack against widespread target systems, complete security of the twilight zone by the enemy is virtually impossible.

f. While guerrillas and the enemy compete for overt control throughout the twilight zone, guerrillas cannot hold any specific area against determined enemy attack. The enemy holds localities which he occupies in force and the guerrillas conduct their operations in those regions where the enemy is weakest.

g. The auxiliary organization is more effective in the twilight zone than it is in enemy-dominated areas. The guerrilla capability of conducting offensive operations coupled with other activities is increased. Intelligence organizations report everything that the enemy does within the twilight zone. Throughout the twilight zone, the enemy is made to feel that lie is in hostile territory; he may control a small segment by force of arms, but lie can never relax his guard lest lie be surprised by guerrillas.

98. Effects

a. Guerrilla operations wear down and inflict casualties upon the enemy, cause damage to supplies and facilities, anti hinder and delay enemy operations. The success of guerrilla operations—even the fact that tile guerrillas continue to exist—lowers enemy morale and prestige; disrupts the economy, politics, and industry of the enemy or enemy occupied areas; and maintains the morale and will to resist of the native population.

b. Because guerrilla operations are primarily directed against lines of communications industrial facilities and key installations, they impede or interdict tile movement of men and materiel and seriously affect the enemy's capability to supply, control, and communicate with his combat forces. In addition, the enemy is compelled to divert manpower and equipment to combat guerrilla activities.

99. Types of Operational Missions

Operational missions for guerrilla forces are categorized as—

a. Missions in Support of the Theater Commander. These missions have their greatest impact on theater level plans. Special forces detachments direct guerrilla forces located ill enemy or enemy occupied territory. Operational command of these unconventional warfare forces is retained by the theater commander and exercised through the SFOB. For a further discussion see paragraph 100.

b. Missions to Assist Conventional Forces Engaged in Combat Operations. These missions are conducted to assist service component tactical commands engaged in combat operations, usually the field army and subordinate elements. Special forces detachments direct guerrilla forces located in enemy occupied territory and operational control of these forces is exercised by the tactical commander through a Special Forces liaison detachment. Logistical and administrative support of unconventional warfare forces remains with the theater commander. For a further discussion see paragraphs 132 through 139.

c. Missions Conducted After Link-Up With Friendly Forces. Missions may be assigned guerrilla forces after link-up with friendly forces has been accomplished. Operational control may be exercised by tactical commanders or passed to other theater army commands such as Theater Army Logistical Command (TALOG) or Theater Army Civil Affairs Command (TACAComd). Special forces detachments may or may not direct the guerrilla force in the execution of these missions. For a further discussion see paragraphs 140 through 147.

100. Missions in Support of the Theater Commander

a. General. These missions may be either strategic or tactical in nature and have both long-range and immediate effects on the enemy and his military forces. They consist of—Interdiction of lines of communications, key areas, military targets, and industrial facilities; psychological operations; special intelligence tasks; and evasion and escape operations.

b. Interdiction. Major emphasis is placed upon interdiction of lines of communications, key areas, industrial facilities, and military targets. Of all guerrilla operations, interdiction usually has the widest impact on the enemy and his ability to wage war and consequently is considered the basic guerrilla warfare operational mission. Interdiction hinders or interrupts the enemy's use of lines of communications, denies him use of certain areas and destroys industrial facilities, military installations, and equipment. Interdiction ranges from simple sabotage by an individual to concerted attacks by guerrilla forces. When properly coordinated with other theater operations, interdiction can make a significant contribution to the overall effectiveness of theater operations. For a discussion of interdiction techniques see paragraphs 101 through 126 and FM 31-21A.

c. Psychological Warfare. All operations are conducted in a manner that will create a favorable environment for psychological control of the indigenous population in keeping with announced postwar objectives. Often the psychological effects of guerrilla operations far outweigh the tactical results. In the operational area, psychological warfare is employed by the area command to communicate with the enemy forces, security forces, active resistance elements and segments of the civilian population supporting, opposing, or indifferent to the resistance movement. Normally, separate psychological operations conducted by guerrilla forces are designed to support the needs of the operational area, and are governed by overall theater objectives. The ability of guerrilla forces to control the population and elicit civilian support is largely dependent upon the psychological impact of the resistance movement upon the populace. For a detailed discussion of psychological operations in support of guerrilla forces, see chapter 9.

d. Special Intelligence Tasks.

(1) Although Special Forces detachments are not intelligence organizations they have the capability through the use of unconventional warfare resources to accomplish certain information gathering. tasks. Intensive intelligence and

reconnaissance activities are conducted to support current and future operations. Such efforts often produce intelligence information of value to other theater forces. Dissemination is made as the situation permits or as directed by SFOB. Specific information collection designed to support other theater forces may be undertaken as directed. These operations are accomplished by either the auxiliary or underground forces under supervision of the area command. Chief among these are target acquisition and damage assessment.

(2) Basic target information can be determined and reported to the SFOB. Because of the tenuous nature of communications between the operational area and the base, target acquisition is usually limited to targets without a high degree of mobility but of vital importance to the theater commander.

(3) Operational detachments can report the physical and psychological effects of attacks conducted by other theater forces against targets within guerrilla warfare areas.

(4) If the importance, magnitude and complexity of intelligence tasks in support of theater commands exceeds the intelligence management capability of unconventional warfare forces, additional intelligence personnel are provided from interested service components. The unconventional warfare force receives these intelligence specialists and furnishes them a base of operations. Although the parent intelligence organizations provide separate communications links for their own personnel, the area commander coordinates their efforts in the interest of security.

e. Evasion and Escape. Evasion and escape mechanisms are developed to assist in the recovery of friendly personnel. Although guerrilla units assist evasion and escape activities, such operations are conducted primarily by auxiliary forces.

Section II. OFFENSIVE COMBAT OPERATIONS

101. General

a. Combat employment of guerrilla forces requires. special forces detachments to direct the efforts of indigenous resistance elements in combat operations. Integrated with these combat operations are psychological warfare, evasion and escape and intelligence activities.

For details of psychological warfare, evasion and escape, and intelligence operations, see chapter 9.

b. Raids and ambushes are the principal offensive techniques of the guerrilla force,, Raids and ambushes may be combined with other action, such as— mining and sniping or these latter actions may be conducted independently. When raids, ambushes, mining and sniping, are directed against enemy lines of communications, key areas, military installations and industrial facilities, the total result is interdiction.

c. Detailed intelligence' of enemy dispositions, movements, and tactics; thorough planning and preparation; and knowledge of the terrain, are prerequisites of guerrilla offensive operations.

102. Characteristics of Guerrilla Combat Operations

Combat operations of guerrilla forces take on certain characteristics which must be understood by Special Forces personnel who direct and coordinate the resistance effort. These characteristics are discussed below.

a. Planning. Careful and detailed planning is a prerequisite for guerrilla combat operations. Plans provide for the attack of selected targets and subsequent operations designed to exploit the advantage gained. Additionally, alternate targets are designated to allow subordinate units a degree of flexibility in taking advantage of sudden changes in the tactical situation. Once committed to an operation the area command has little capability to rapidly manipulate subordinate units to other missions, This lack of immediate response is due to the shortage or non-existence of radio communications equipment within smaller guerrilla units coupled with relatively large zones of action. Thus, plans must be thorough and flexible enough to allow commanders who are responsible for an operation or series of operations to adopt alternate predetermined courses of action when contingencies arise.

b. Intelligence. The basis of planning is accurate and up-to-date intelligence. Prior to initiating combat operations, a detailed intelligence collection effort is made in the projected objective area. This effort supplements the regular flow of intelligence. Provisions are made for keeping the target or objective area under surveillance up to the time of attack.

c. Decentralized Execution. Guerrilla combat operations feature centralized planning and decentralized execution. Action of all resistance elements is directed and coordinated by the area command. However, within the guidance furnished by the area commander, subordinate units are allowed the widest possible latitude in the conduct of operations.

d. Surprise. Guerrilla combat, operations stress surprise. Attacks are executed at unexpected times and places. Set patterns of action are avoided. Maximum advantage is gained by attacking enemy weaknesses. Low visibility and adverse weather are exploited by guerrilla forces. Surprise may also be enhanced by the conduct of concurrent diversionary activities.

e. Short Duration Action. Usually, combat operations of guerrilla forces are marked by action of short duration against the target followed by a rapid withdrawal of the attacking force. Prolonged combat action from fixed positions is avoided,

f. Multiple Attacks (fig. 25). Another characteristic of guerrilla combat operations is the employment of multiple attacks over a wide area by small units tailored to the individual mission. This is not piecemeal commitment of units against single targets but a number of attacks directed against several targets or portions of the target system. Such action tends to deceive the enemy as to the actual location of guerrilla bases, causes him to over-estimate guerrilla strength and forces him to disperse his rear area security and counter guerrilla efforts.

103. Tactical Control Measures

a. General. The area commander utilizes tactical control measures to aid him in directing and coordinating combat operations. Common tactical control measures are—

(1) Targets (objectives).

(2) Zones of action.

(3) Axis of advance.

(4) Mission support sites.

b. Target (Objectives). The area commander designates targets or objectives for attack by subordinate units. These targets are usually lines of communications, military installations and units and industrial facilities. Normally, targets or objectives for guerrilla forces are not held for any length of time nor are they cleared of determined enemy resistance.

c. Zones of Action (fig. 26). Zones of action are used to designate areas of responsibility for operations of subordinate units. Within the zone of action the subordinate commander exercises considerable freedom in the conduct of operations. Movement of other guerrilla units through an adjacent zone of action is coordinated by the area command. The auxiliary forces within a zone of action provide support to the guerrilla unit responsible for the area. Boundaries of

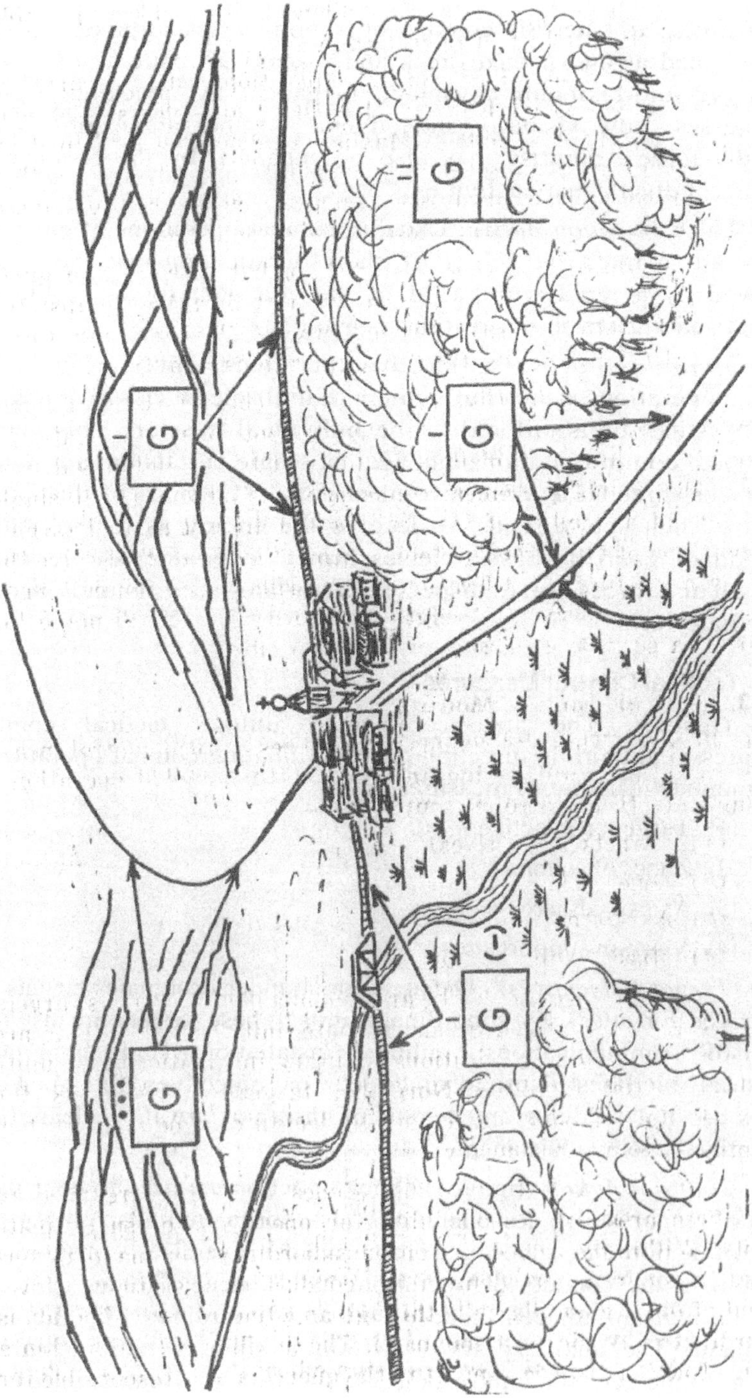

Figure 25. Multiple attacks by guerrilla units.

zones of action are changed by the commander who established them as required.

d. Axis of Advance. Guerrilla commanders may prescribe axes of advance for their unit or subordinate units in order to control movement to targets. Guerrilla units move to the objective area either by single or multiple routes.

e. Mission Support Sites (fig. 27). Mission support sites are utilized by guerrilla units to add reach to their operations and enable them to remain away from guerrilla bases for longer periods of time. The mission support site is a pre-selected area used as a temporary stopover point and is located in areas not controlled by the guerrilla force. Mission support sites are utilized prior to and/or after an operation. They are occupied for short periods of time, seldom longer than a day. As in an assembly area, the using unit prepares for further operations and may be provided with supplies and intelligence by auxiliary forces.

f. Additional Tactical Control Measures. Additional control measures may be employed by smaller guerrilla units such as rallying points, direction of attack, assault positions and lines of departure. These control measures are employed in a manner similar to their use by conventional military units.

104. Target Selection

a. The general mission assigned by the theater commander determines the type target (objective) to be attacked, with the final selection of the specific target usually made by the detachment commander. Occasionally, the SFOB may select the target. The important factors related to the target which influence its final selection are

(1) Criticality. A target is critical when its destruction or damage will exercise a significant influence upon the enemy's ability to conduct or support operations. Such targets as bridges, tunnels, ravines, and mountain passes are critical to lines of communication; engines, tires, and POL stores are critical to transportation. Each target is considered in relationship to other elements of the target system.

(2) Vulnerability. Vulnerability is a target's susceptibility to attack by means available to UW forces. Vulnerability is influenced by the nature of the target, i.e., type, size, disposition and composition.

(3) Accessibility. Accessibility is measured by the ability of the attacker to infiltrate into the target area. In studying a target for accessibility, security controls around the target

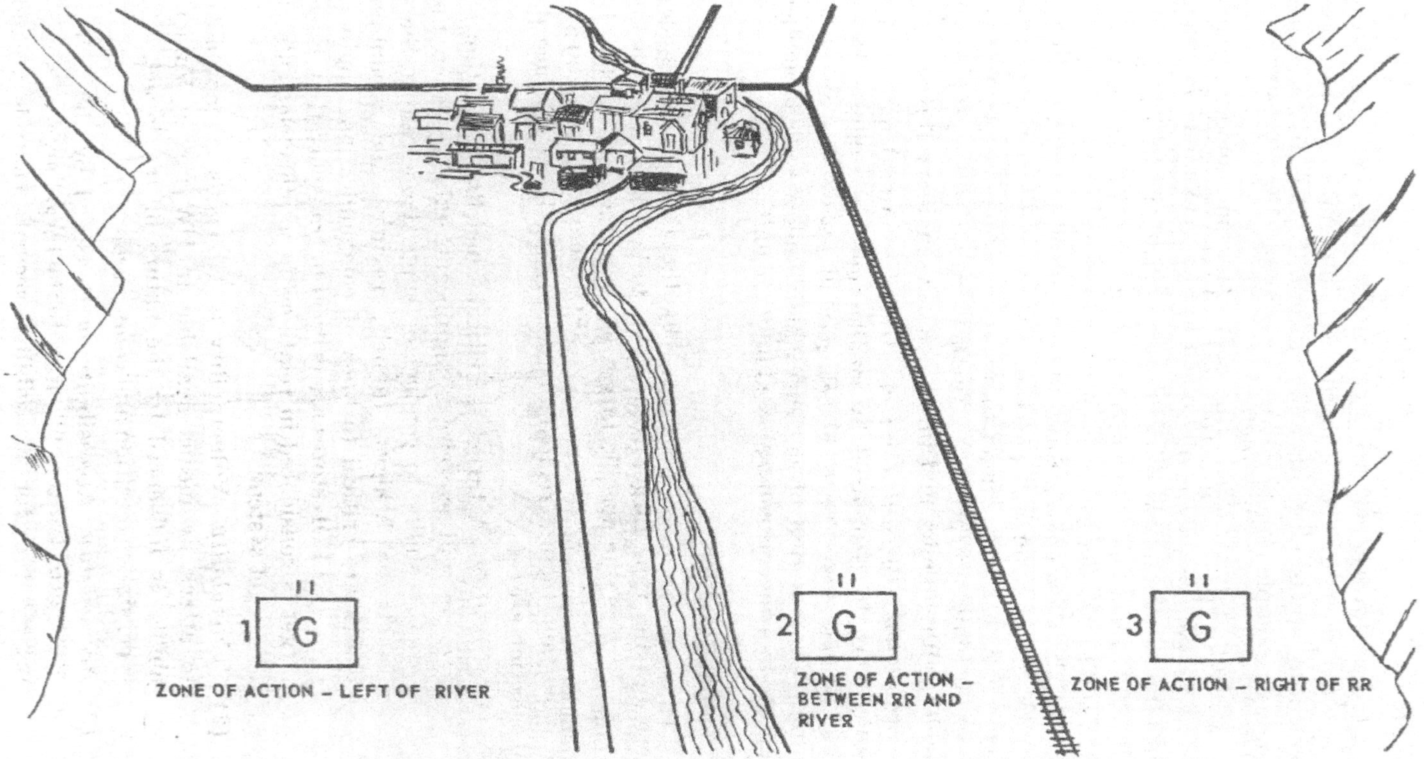

Figure 26. Tactical control measures—zones of action.

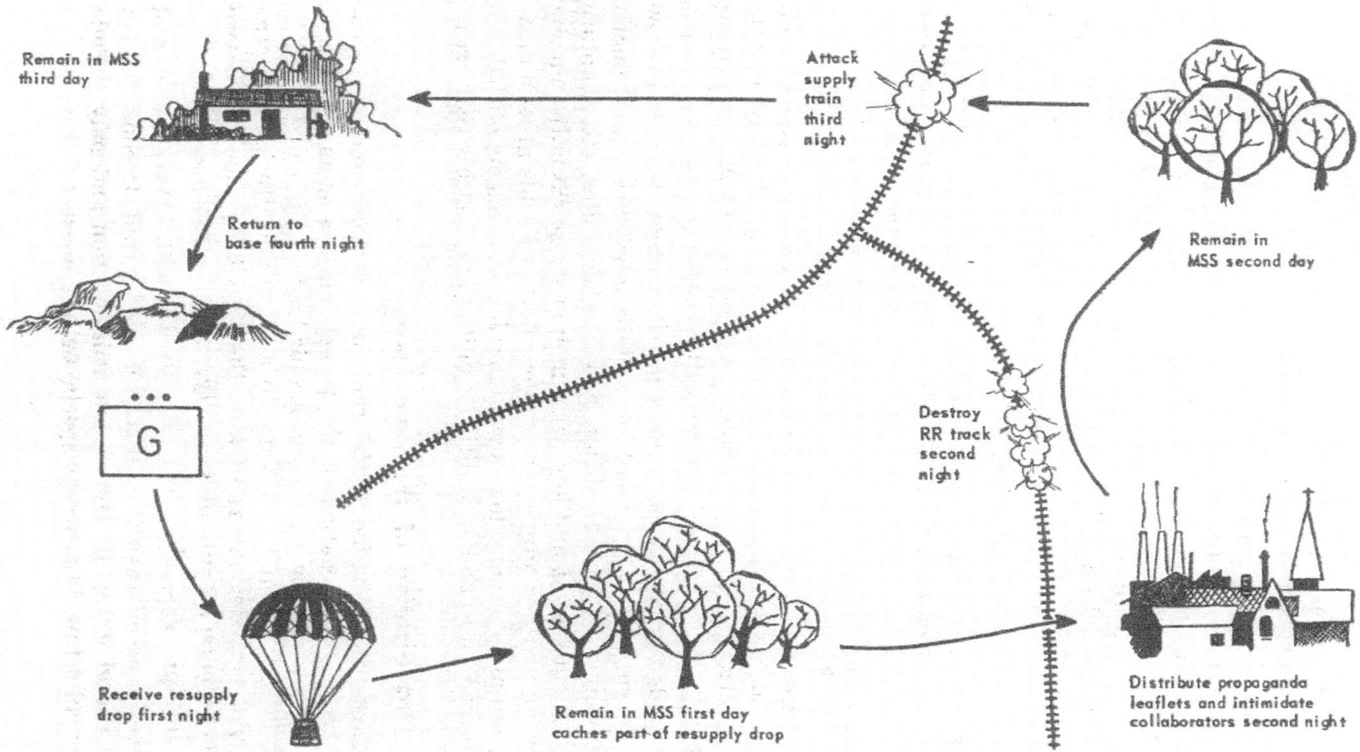

Figure 27. Tactical control measures—mission support site.

113

area, location of the target, and means of infiltration are considered.

(4) *Recuperability.* Recuperability is the enemy's ability to restore a damaged facility to normal operating capacity. It is affected by the enemy capability to repair and replace damaged portions of the target.

b. These factors are variables. As such, each target is constantly reevaluated for selection. The criticality of a target may change from time to time. A railroad bridge is less critical when the enemy has few locomotives. The vulnerability of a target shifts with the means available to attack it such as explosives, incendiaries and special devices. A power plant, command post, or supply depot is less accessible after the enemy has detailed additional security personnel to guard it. Recuperation is more certain if reserve stocks are plentiful.

c. Each operation plan includes more than one target. This enables the commander to alter his plans to meet unforeseen situations that may preclude attack of the primary target.

105. Raids

a. A raid is a surprise attack against an enemy force or installation. Such attacks are characterized by secret movement to the objective area; brief, violent combat; rapid disengagement from action; and swift, deceptive withdrawal.

b. Raids are conducted by guerrilla units to—destroy or damage supplies, equipment or installations such as command posts, communication facilities, depots, radar sites, etc.; capture supplies, equipment and key personnel; or cause casualties among the enemy and his supporters. Other effects of raids are: to draw attention away from other operations; keep the enemy off balance and force him to deploy additional units to protect his rear areas.

106. Organization of the Raid Force

a. General. The size of the raid force depends upon the mission, nature and location of the target and the enemy situation. The raid force may vary from a squad attacking a police checkpoint or unprotected rail lines, to a battalion attacking a large supply depot. Regardless of size, the raid force consists of two basic elements; assault and security.

b. Assault Element. The assault element is organized and trained to accomplish the objectives of the raid. It consists of a main action group to execute the raid mission and may include personnel detailed to execute special tasks

(1) The main action group executes the major task, the accomplishment of which insures the success of the raid. For instance, if the raid objective is to destroy a critical installation such as a railroad bridge or tunnel, the main action group emplaces and detonates the demolition charges. In the event that the target can be neutralized by fire, such as enemy personnel, the main action group conducts its attack with a high proportion of automatic weapons. In some instances the main action group moves physically on or into the target; in others they are able to accomplish their task at a distance from the target. The efforts of other ,elements of the raid force are designed to allow the main action group access to the target for the time required to accomplish the raid mission.

(2) If required, special task details assist the main action group to reach the target. They execute such complementary tasks as• eliminating guards, breaching and removing obstacles, diversionary or holding tasks, and fire support. The special task details may precede, act concurrently with, or follow the main action group.

c. Security Element. The security element supports the raid by preventing the enemy from reinforcing or escaping. Additionally, the security element covers the withdrawal of the assault element and acts as a rear guard for the raid force. The size of the security element depends upon the enemy's, capability to intervene in the operation.

107. Preparation

a. Planning Considerations.

(1) The first step is the selection of a target. In addition to the target selection factors mentioned in paragraph 104, other important considerations are in the nature of the terrain and the combat efficiency of the guerrilla force.

(2) Additionally, the UW force commander must consider possible adverse effects on his unit and the civilian populace. The guerrillas' objective is to diminish the enemy's military potential, but an improperly timed operation may provoke counteraction which they are unprepared to meet. Successful operations increase guerrilla prestige with the civilians and make them more

willing to provide support. However, the guerrillas should take every precaution to insure that civilians are not needlessly subjected to harsh reprisals. Success also enhances the morale of the guerrilla unit and increases the prestige of its leaders. On the other hand, an unsuccessful attack often has disastrous effects on guerrilla morale.

(3) Although detailed, the plan for a raid must be essentially simple, and not depend on too many contingencies for its success. Duplicate or alternate arrangements are made for the execution of key operations to increase the chances of success. Guerrilla activities in the area are planned so as to give the installation no indication of the pending attack; This may either be suspension, continuation or increase of current activity. Time and space factors are carefully considered when planning the operation. Sufficient time is allowed for assembly and movement, particularly during darkness; the require menu of the situation determine whether movement and attack should be made during daylight or darkness. Darkness favors surprise and is usually the best time when the operation is simple and the physical arrangement of the installation is well known, Early dawn or dusk is favored when inadequate knowledge of the installation or other factors necessitate close control of the operation. A withdrawal late in the day or at night makes close pursuit by the enemy more difficult.

b. Intelligence. The raid force commander must have maximum available intelligence of—the target; enemy forces able to intervene; the civilian population in the vicinity of the target; and the terrain to be traversed en route to and returning from the objective area. An intensive intelligence effort precedes the raid. guerrilla intelligence and reconnaissance elements conduct reconnaissance of the routes to the target and if possible of the target itself. Local auxiliary sources are exploited and the auxiliaries may be required to furnish guides. Surveillance of the target is continuous up to the time of the attack. The raid force commander exercises extreme caution to deny the enemy any indications of the impending operation through action of either guerrilla reconnaissance elements or the auxiliaries,

c. Rehearsals of Participants. All participants are rehearsed for the operation. Rehearsals ax a conducted as realistically as possible. If available, terrain—similar to that found in the target area is used. The use of sand tables, sketches, photographs, and target mockups assist in the briefing of guerrilla troops. Contingency actions aye also practiced. Final rehearsals are conducted under conditions of visibility like those expected in the objective area.

d. Final Inspection. The raid force commander conducts a final inspection of personnel and equipment prior to movement to the objective area. Weapons are test fired if possible, faulty equipment is replaced and the condition of the men is checked, During this inspection a counterintelligence check is made of each guerrilla's personal belongings to insure that no incriminating documents are carried during the operation. This inspection assures the raid force commander that his unit is equipped and conditioned for the operation.

108. Movement

(fig. 28)

Movement to the objective area is planned and conducted to allow the raid force to approach the target undetected. Movement may be over single or multiple routes. The preselected route or routes terminates in or near one or more mission support sites, During movement every effort is made to avoid contact with the enemy, Upon reaching the mission support site, security groups are deployed and final coordination takes place prior to movement to the attack position.

109. Action in the Objective Area

(fig. 29)

Special task details move to them positions and 'eliminate sentries, breach or remove obstacles and execute other assigned tasks. The main action group quickly Toll• ows the special task details into the target area. Once the objective of the raid has been accomplished the main action group withdraws covered by designated fire support elements and/or dart of the security force. In the event the attack is unsuccessful the action is terminated to prevent undue loss. Special task details withdraw according to plan. The assault element may assemble; at one or more rallying points. The security elements remain in position to cover 'the withdrawal of the assault elements and withdraw on signal or at a prearranged time.

110. Withdrawal

(fig. 30)

a. Withdrawal is accomplished in a manner designed to achieve maximum deception of the enemy and to facilitate further action by the raid force: The various elements of the raiding force withdraw, in order, over predetermined routes through a series of rallying points:

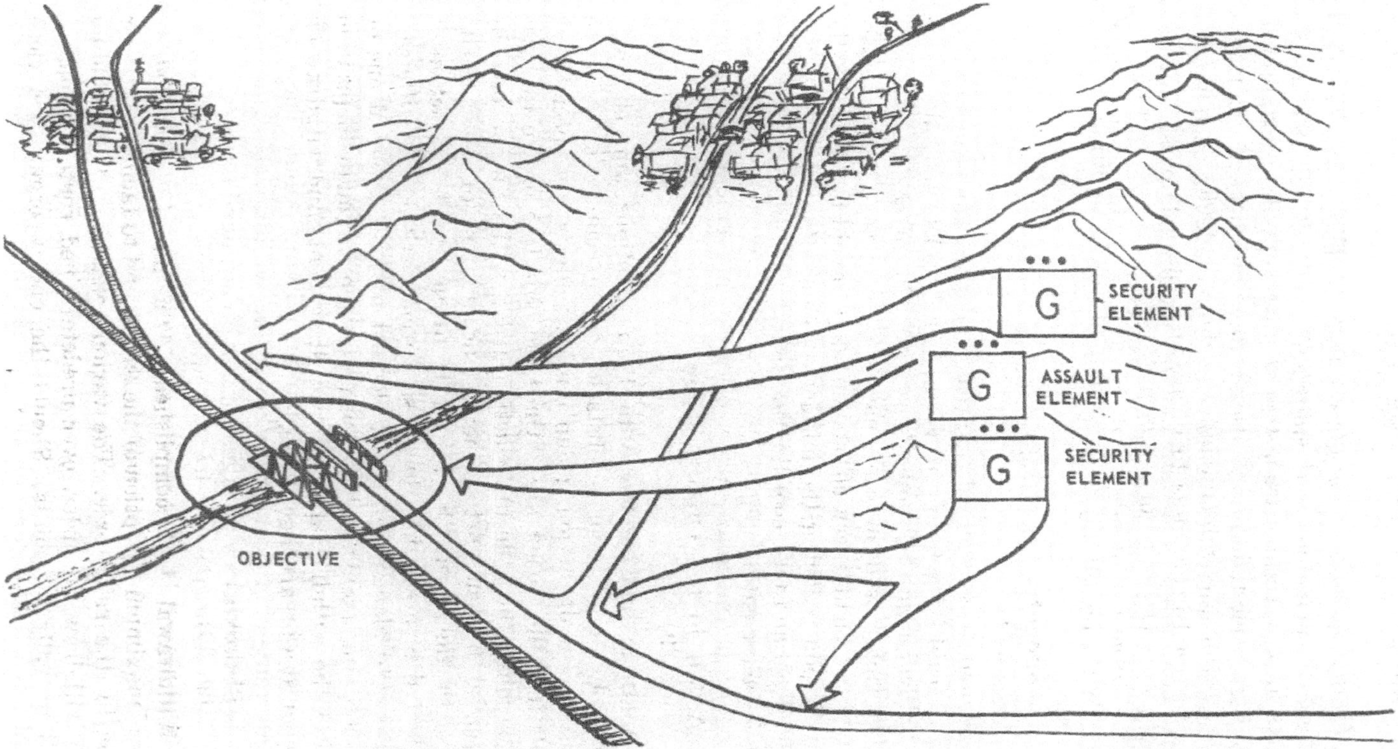

Figure 28. Raid—movement to the objective.

Figure 29. Raid—action in the objective area.

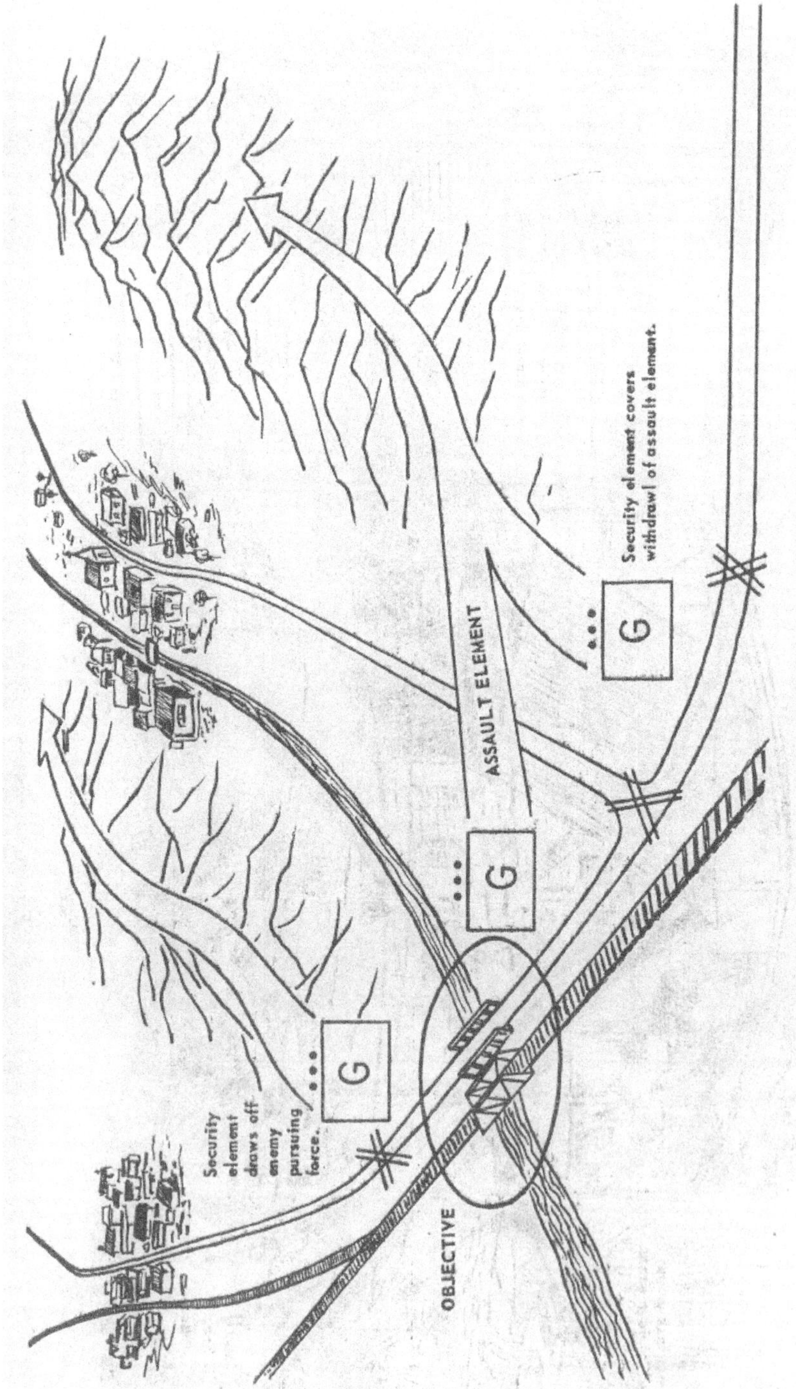

Figure 30. Raid—withdrawal from action.

Should the enemy organize a close pursuit of the assault element, the security element assists by fire and movement, distracting the enemy and slowing him down. Elements of the raiding force which are closely pursued by the enemy do not attempt to reach the initial rallying point, but on their own initiative lead the enemy away from the remainder of the force and attempt to lose him by evasive action over difficult terrain. If the situation permits, an attempt is made to reestablish contact with the raid force at other rallying points or to continue to the base area as a separate group. When necessary, the raiding force, or elements of it, separate into small groups or even individuals to wade close pursuit by the enemy.

b. Frequently, the raid force disperses into smaller units, withdraws in different directions and reassembles at a later time and at a predesignated place to conduct other operations. Elements of the raid force may conduct further operations, such as an ambush of the pursuing enemy force, during the withdrawal.

111. Large Raid

(fig, 31)

a. General. When a target is large, important to the enemy, and well guarded, the size of the guerrilla force required to effectively attack it increases. Large raids involve the use of a battalion or more of guerrillas. Essentially the operation is conducted similar to smaller raids, however, additional problems must be considered.

b. Movement to Objective Area. Surprise is as desirable • in a large as well as a smaller raid but is usually harder to achieve. The numbers of troops to be deployed requires additional mission support sites. Again the size of the guerrilla force may require selection of mission support sites at a greater distance from the target to preserve secrecy, thus requiring a longer move to the attack position. A large raid force usually moves by small components over multiple routes to the objective area,

c. Control. Another problem inherent in a large raid is that of control. Guerrilla units without extensive radio communications equipment will find coordination of various widespread elements is difficult to achieve. Pyrotechnics, audible signals, runners or predesignated times may be used to coordinate action.

d. Training. A high degree of training and discipline is required to execute a large raid. Extensive rehearsals assist in preparing the force for the mission. In particular commanders and staffs must learn to employ the larger number of troops as a cohesive force.

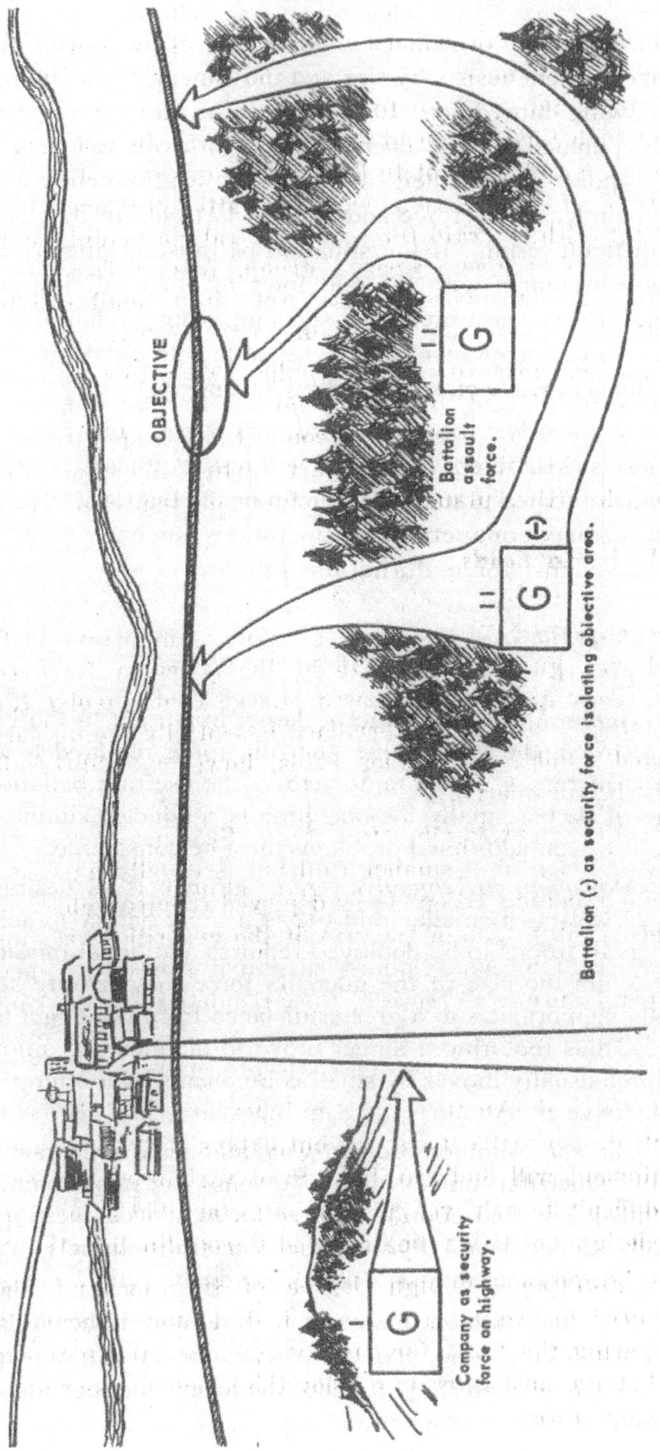

Figure 31. Large raid.

e. Fire Support. Additional fire support is usually a requirement. This may mean secretly caching ammunition in mission support sites over a period of time prior to the raid. Guerrillas may each carry a mortar or recoilless rifle round, rocket or box of machinegun ammunition and leave them at a mission support site or firing position for fire support units.

f. Timing. Timing is usually more difficult for a large raid. The time to move units and time the main action element needs to perform its mission are usually longer. This requires stronger security elements to isolate the objective area for longer periods. The timing of the raid takes on increased importance because of the large numbers of guerrillas involved. Movement to the objective area is usually accomplished during periods of low visibility, however because of fire support coordination requirements and larger numbers of personnel, the action may take place during daylight hours.

g. Withdrawal. Withdrawal from a large raid is usually by smaller groups over multiple routes in order to deceive the enemy and dissipate his pursuit. Dispersed withdrawal has the added advantage of not providing a lucrative target to enemy air and fire support elements. However, the raid force commander must consider the possibility of defeat in detail of his force by an alert and aggressive enemy. The decision as to how to conduct the withdrawal must be based on a careful weighing of these factors.

112. Ambushes

a. An ambush is a surprise attack used against moving or temporarily halted targets such as railroad trains, truck convoys, individual vehicles, and dismounted troops. In an ambush, the enemy sets the time and the attacker, the place.

b. Ambushes are conducted to destroy or capture personnel and supplies; harass and demoralize the enemy; delay or block movement of personnel and supplies; and canalize enemy movement by making certain routes useless for traffic. The result usually is concentration of the majority of movements to principal roads and railroads where targets are more vulnerable to attack by other theater forces.

c. Like the raid force, the ambush force is organized into assault and security elements. The assault element conducts the main attack against the ambush target which includes halting the column, killing or capturing personnel, recovering supplies and equipment and destroying unwanted vehicles or supplies which cannot be moved. The security force isolates the ambush site Using roadblocks, other ambushes and outposts, Security elements cover the withdrawal of the assault element.

113. Preparation

Preparation for an ambush is similar to that of a raid except that selection of the ambush site is an additional consideration,

a. Planning Considerations. Begin preparing the ambush plan, consideration is liven to—

(1) The mission—this may be, a single ambush against one column or a series of ambushes against one pr more routes of communication.

(2) The probable sizes strength, end composition of the enemy force that is to be ambushed; formations likely to be used, and his reinforcement capability.

(3) Terrain along the route favorable for an ambush, including unobserved routes, of approach and withdrawal.

(4) Timing of the ambush—ambushes conducted during periods of low visibility offer a wider choir 2 of positions and better opportunities to surprise and confuse the enemy than daylight ambushes. However, control and movement to and during the night ambush is more difficult. Night ambushes are more suitable when the mission can be accomplished during or immediately following the initial burst of fire. They require a maxi, mum number of automatic weapons to be used at close range, Night ambushes are effective in hindering the enemy's use of routes of communication by night, while friendly aircraft attack the same routes during the day. Daylight ambushes facilitate control and permit offensive action for a longer period of time. A day ambush also provides opportunity for more effective aimed fire of such weapons as rocket launchers and recoilless rifles.

b. Intelligence. Since the guerrillas are seldom able to ascertain in advance the exact composition, strength and time of movement of convoys, their intelligence effort should be directed towards determining the convoy pattern of the enemy. Using this information, guerrilla commanders are able to decide on. type convoys to be attacked by ambush. In addition, intelligence considerations described in paragraph 107 for R raid are equally applicable to an ambush.

c. Site Selection. In selecting the ambush site, the basic consideration is favorable terrain, although limitations which may exist such as deficiencies in the firepower of guerrillas and lack of resupply during actions may govern the choice of ambush site. The site should have firing positions which offer concealment and favorable fields of fire. Whenever possible, firing should be done through a screen of foliage. The terrain at the site should serve to funnel the enemy into a killing zone. The entire killing zone is covered by fire to avoid dead space that would allow the enemy to organize resistance. The guerrilla force should take advantage of natural obstacles such as defiles, swamps, and cliffs which will restrict enemy maneuver against the ambush force. When natural obstacles do not exist, mines and demolitions are employed to canalize the enemy. Security elements are placed on roads and trails leading to the ambush site to warn the assault element of the enemy approach. These security elements also assist in covering the withdrawal of the assault element from the ambush site. The proximity of security to assault elements is dictated by the terrain. In many instances, it may be necessary to organize secondary ambushes and roadblocks to intercept and delay enemy reinforcements.

114. Conduct of the Ambush

a. Movement. The guerrilla force moves over a, preselected route or routes to the ambush site. One or more mission support sites are usually necessary along the route to the ambush site. Last minute intelligence is provided by reconnaissance elements and final coordination for the ambush is made at the mission support site.

b. Action at the Ambush Site (fig. 32).

(1) Troops are moved to an assembly area near the ambush site and security elements take their positions. As the approaching enemy column is detected, or at a predesignated time, the ambush commander decides whether or not to execute the ambush. This decision depends on size of the column, guard and security measures, and estimated worth of the target in light of the mission. If a decision is made to execute the ambush, advance guards are allowed to pass through the main position. When the head of the main column reaches a predetermined point, it is halted by fire, demolitions, or obstacles, At this signal, the entire assault element opens fire, Designated details engage the advance and rear guards to prevent reinforcement of the main column. The volume of fire is rapid and directed at enemy personnel, exist from vehicles, and automatic weapons. Anti-tank grenades, rocket launchers, and recoilless rifles are used against armored vehicles. Machine-guns lay bands of fixed

fire across escape routes. Mortar shells, hand and rifle grenades are fired into the killing zone. If the commander decides to assault, it is launched under covering fire on a prearranged signal. After enemy resistance has been nullified special parties move into the column to recover supplies, equipment and ammunition. When the commander desires to terminate the action because either the mission has been accomplished, or superior enemy reinforcements are arriving, he withdraws first the assault element and then the security elements. The security elements cover the withdrawal of the assault element.

(2) If the purpose of the ambush is to harass and demoralize the enemy a different approach may be adopted. The advance guard is selected as the target of the ambush and the fire of the assault element is directed against them. Repeated attacks against enemy advance guards have the following effects:

(a) They cause him to use disproportionately strong forces in advance guard duties. This may leave other portions of the column vulnerable or require him to divert additional troops to convoy duty.

(b) They have an adverse psychological effect upon enemy troops. Continued casualties incurred by the advance guard make such duty unpopular.

c. Withdrawal. Withdrawal from the ambush site is covered by the security elements in a manner similar to the withdrawal from a raid (par. 110).

115. Special Ambush Situations

a. Columns Protected by Armor. Attacks against columns protected by armored vehicles depend upon the type and location of armored vehicles in a column and the weapons of the ambush force. If possible, armored vehicles are destroyed or disabled by fire of anti-tank weapons, land mines, molotov cocktails, or by throwing hand grenades into open hatches. An effort is made to immobilize armored vehicles at a point where they are unable to give protection to the rest of the convoy and block the route of other supporting vehicles.

b. Ambush of Railroad Trains. Moving trains may be subjected to harassing fire, but the most effective ambush involves derailing the train. The locomotive should be derailed on a down grade, at a sharp curve or on a high bridge. This causes most of the cars to overturn and results in extensive casualties among passengers. It is desirable

Figure 32. Action at the ambush site.

to derail trains so that the wreckage remains on the tracks to delay traffic for longer periods of time. Fire is directed on the exits of overturned coaches and designated groups armed with automatic weapons rush forward to assault coaches which are still standing. Other groups take supplies from freight cars and then set fire to the train. Rails are removed from the track at some distance from the ambush site in each direction to delay the arrival of reinforcements by train. In planning the ambush of. a train, remember that the enemy may include armored railroad cars in the train for its protection and that important trains may be preceded by advance guard locomotives or inspection cars to check the track.

c. Ambush of Waterway Traffic. Waterway traffic like barges, ships, and other craft may be ambushed in a manner similar to a vehicular column. The ambush party may be able to mine the waterway and thus stop traffic. If mining is not feasible, fire delivered by recoilless weapons can damage or sink the craft. Fire should be directed at engine room spaces, the waterline and the bridge. Recovery of supplies may be possible if the craft is beached on the banks of the waterway or grounded in shallow water.

116. Mining and Sniping

a. Mining (fig. 33) .

(1) Mining affords the area commander a means of interdicting enemy routes of communication and key areas with little expenditure of manpower. Additionally, mines allow the user to move away from the mined site before the enemy activates them. The planned use of mines as an interdiction technique also has a demoralizing effect on enemy morale.

(2) Mines may be employed in conjunction with other operations, such as raids, ambushes and sniping, or used alone. When utilized alone they are emplaced along routes of communication or known enemy approaches within an area at a time when traffic is light. This allows personnel emplacing the mines to complete the task without undue interference and then make good their escape,

(3) The use of mines to' cover the withdrawal of a raiding or ambush force slows enemy pursuit. Their utilization in roadbeds of highways and railroads interferes with movement. Mines may be emplaced around enemy installations. These mines will cause casualties to sentinels

Enemy sentinels activate mine
placed by guerrillas.

and patrols and tend to limit movement outside of enemy installations.

b. Sniping (fig. 34). Sniping is an interdiction technique. It is economical in the use of personnel and has a demoralizing effect on enemy forces. A few trained snipers can cause casualties among enemy personnel, deny or .hinder his use of certain routes and require him to employ a disproportionate number of troops to drive off the snipers. Snipers may operate to cover a mined area, as part of a raiding or ambush force or by themselves. Snipers operate best in teams of two, alternating the duties of observer and sniper between themselves.

Section III. INTERDICTION

117. General

a. UW forces use interdiction as the primary means of accomplishing operational objectives. Interdiction is designed to prevent or hinder, by any means, enemy use of an area or route. Interdiction is the cumulative effect of numerous smaller offensive operations such as raids, ambushes, mining, and sniping. Enemy areas or routes that offer the most vulnerable and lucrative targets for interdiction are industrial facilities, military installations, and lines of communication.

b. The results of planned interdiction programs are

(1) Effective interference with the movement of personnel, supplies, equipment and raw material.

(2) Destruction of storage and production facilities.

(3) Destruction of Military installations. For positive results attacks are directed against the primary and alternate critical elements of each target system.

c. Profitable secondary results can be obtained from interdiction operations if they are conducted over a large area. When the UW force employs units in rapid attacks in different and widely spaced places it:

(1) Makes it difficult for the enemy to accurately locate guerrilla bases by analyzing guerrilla operations.

(2) Causes the enemy to over-estimate the strength and support of the guerrilla force.

(3) May tend to demoralize him and lessen his will to fight.

d. Suitable targets for interdiction are facilities and material utilized by an enemy to support his war effort. Major targets susceptible to attack by UW forces include:

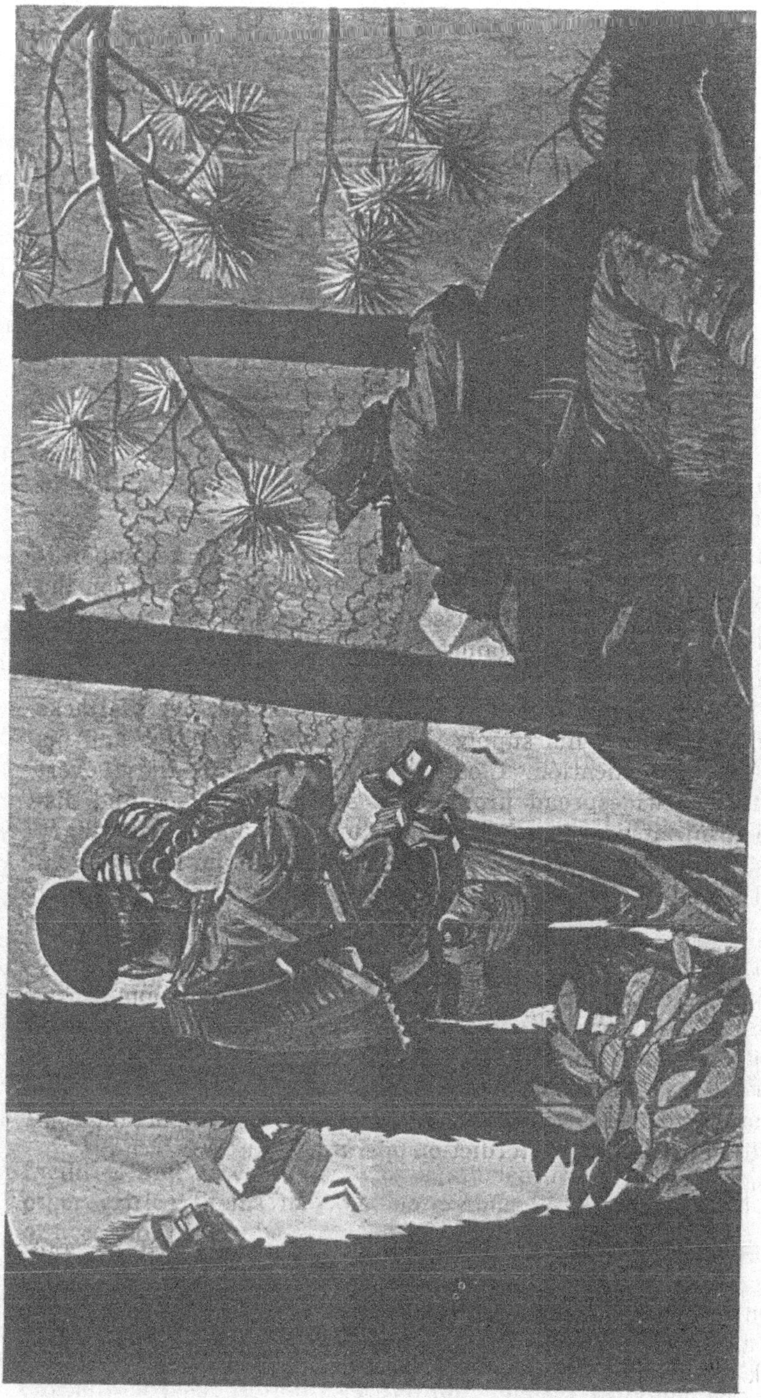

Figure 34. Sniping.

(1) Transportation—railroad, highway, water, air.

(2) Communication—telephone, telegraph, radio, television.

(3) Industry—manufacturing facilities for weapons, aircraft, vehicles, ammunition, shipping, etc.

(4) Power—electric, nuclear, chemical.

(5) Fuel—gas, oil.

(6) Military installations and personnel.

118. Planning

a. The UW force commander bases interdiction planning upon his mission. The mission should specify the results desired by the higher commander in an operational area and prescribes priorities of attack against target systems. The result of interdiction by UW elements combined with attacks conducted by other forces is designed to seriously hamper or destroy the enemy's ability to support his combat forces.

b. The area commander selects targets and designates subordinate elements to attack them. Target selection is based upon the mission and the criteria discussed in paragraph 104. Normally, operations are directed against targets on as broad a scale as possible utilizing all available UW elements which have a capability to attack the target. Guerrilla units conduct overt attacks against the enemy, his supply aid production facilities and his lines of communication. Closely coordinated with these overt attacks—is a widespread program of sabotage, strikes and disaffection initiated and directed by the underground and auxiliary forces. Attacks are timed so as to achieve maximum results from surprise and confusion and often coincide with operations of other theater forces.

c. The enemy reaction to wide scale UW operations is usually violent, immediate and directed against the civilian population. Inevitably, losses among civilian support elements (auxiliaries and the underground) may be high. Continued pressure by the area command may lessen or divert this reaction to other areas. The effects of enemy reaction on the UW force is an important consideration in planning interdiction operations.

119. Railroad Systems

a. General. Railroads presents one of the most profitable and easily accessible target systems for attack by guerrilla forces. In general, open stretches of track; switches, repair facilities, and coal and water supplies provide unlimited opportunities for attack. On electrified railroads, power, sub-stations, plants end lines offer critical targets. Types of railway targets vary with the geographical area.

b. Tracks. Railroad tracks are easily attacked by guerrilla units because it is almost impossible to guard long stretches of track effectively. Lightly armed, mobile guerrilla units can inflict heavy damage on tracks. Guerrilla attacks against rails have far reaching effects on an enemy who relies heavily upon railroad traffic for military movement.

(1) Attacks on open tracks use fewer explosives than attacks on other railroad installations. an eight- to ten-man guerrilla unit can destroy a considerable amount of railroad track in a night. It is possible for a small group of guerrillas working regularly to keep a single track out of operation permanently.

(2) Attacks on, tracks should cover a wide area. Multiple breaks should be made in areas in which guerrilla forces of squad size or larger can be used. Single breaks by individuals or very small teams should be made on a large perimeter and in areas not accessible to larger guerrilla forces. Telegraph and telephone lines along the railroad are cut simultaneously.

(3) When conducting attacks on more than one railroad line, attacks are carefully planned to use guerrilla forces and supplies economically and to the best advantage. The determination of main arteries of railroad traffic and their connecting lines is essential, in planning for attacks against a rail system.

(4) When necessary, security elements are placed on the flanks of the attacking elements, along the tracks, and on any roads leading to the target area. Coordination is made, particularly at night, so that small units attacking a stretch of rail line do not become accidentally engaged in fire fights among themselves. Successive rallying points are designated to permit withdrawal of units fir reorganization.

c. Critical Equipment. Because they are usually guarded, repair facilities and reserve stocks of equipment, railroad cranes, and other critical items may be more difficult to attack. This lack of accessibility can be overcome by carefully planned and executed operations. Results expected from these operations are weighed against the probability of increased guerrilla casualties.

d. Rolling Stock. Rolling stock may be simultaneously attacked with track interdiction. Demolition of tracks, at the time when trains are passing can increase the damage to the tracks and track bed, result in captured supplies, kill and wound enemy personnel, or liberate prisoners. Trains moving through areas menaced by guerrillas move slowly and are guarded. Attacks on guarded trains require well-trained and well-armed guerrillas. Rocket launchers or other weapons capable of firing large caliber AP ammunition are usually necessary; mines may also be used.

e. Critical Points. Bridges, tunnels, and narrow railway passes are usually well-guarded. Repair equipment and bridging equipment are normally located in the vicinity, and should be attacked concurrently.

f. Effect of Railway Interdiction. Limited operations against tracks and traffic only cause harassment, therefore widespread operations are necessary before any severe effect is felt by the enemy. Harassment of repair crews by snipers and ambushes is effective in reducing enemy morale and the willingness of his personnel to participate in repair work.

(1) Underground and auxiliary units interdict railroads in areas away from guerrilla control.

(2) Interdiction of rail traffic over a wide area is usually more effective than attacks aimed at complete destruction of a short stretch of railroad. Apart from the psychological effect on the enemy forces and civilian population, interdiction over a wide area hampers enemy movement more than limited total destruction.

(3) The early interdiction of railroads interferes with the enemy's offensive momentum and may forestall largescale deportation of civilian populations. The primary effect of interdiction of railroads is disruption of the enemy's flow of supplies, movement of troops, and industrial production. Secondary effects are—

(a) Disruption of the orderly processes of dispatching and controlling rail movements, which in turn may result in the accumulation of sizeable targets at rail terminals, junctions and marshalling yards. These targets are then susceptible to attack by other service components.

(b) Depletion of reserves of repair materials which often results in the dismantling of secondary rail lines for the repair of primary lines.

(c) Transfer of rail traffic to primary roads and highways, which are vulnerable to guerrilla and air attack.

(d) Increasing the burden upon enemy security forces and repair crews.

120. Highway Systems

a. Highways are less vulnerable targets than railroads. Damage inflicted is more easily repaired and repairs require fewer critical materials and less skilled labor.

b. Bridges, underpasses, and tunnels are vulnerable points on road networks. Sections of road which may be destroyed by flooding from adjacent rivers, canals or lakes are also vulnerable. In addition, a road may be interdicted by causing rock or land slides.

c. Since highways have fewer vulnerable spots, it is likely that these points will be heavily defended. This requires a large guerrilla force and the use of heavier weapons to neutralize protecting pillboxes and fortifications. Because of this, it is better to concentrate on attacking enemy convoys and columns using the highways. In the initial stages of hostilities, small bridges, tunnels, cuts, culverts and levees may be insufficiently protected. As guerrilla attacks increase in frequency and effect, enemy security forces increase protection of these likely guerrilla targets.

d. Where the roads cannot be destroyed, traffic is interrupted by real and dummy mines. Ambushes are conducted when suitable terrain is available. Long-range fires from positions away from roads disrupt enemy traffic.

e. Points for interdiction are selected in areas where the enemy cannot easily re-establish movement by making a short detour.

121. Waterway Systems

a. The vulnerable portions of waterway systems are electrical installations, dams and locks which are usually well guarded. The destruction of these installations can disrupt traffic effectively for long periods. other waterway installations such as signal lights, beacons and channel markers can be effectively attacked. Sinking vessels in restricted channels by floating mines, limpets, or fire from heavy caliber weapons may be effective in blocking waterway traffic.

b. Dropping bridges into the waterway, creating slides, and destroying levees all hinder ship movement on waterways.

c. Personnel who operate the waterway facilities such as pilots and lock operators may be eliminated. These personnel are not easily replaced and their loss will effect operation of the waterway.

d. Mines and demolitions charges may be placed at strategic points on the waterway. If floating mines are used the waterway is reconnoitered for possible anti-mine nets. Cables supporting these nets are attached to poles or trees on both banks of the waterway or are supported by boats in the stream and should be cut before employing floating mines.

122. Airways Systems

a. Airways are interdicted by attacking those facilities that support air movement. Air terminals, communications systems, navigational systems, POL dumps, maintenance facilities and key personnel are targets for attack.

b. Since air traffic is dependent upon fuel, lubricants, spare parts and maintenance tools, lines of communications and installations providing these items are attacked.

123. Communication Systems

a. Wire communications are vulnerable to guerrilla attack, however, destruction of a single axis of a wire system seldom results in the complete loss of long distance telephone or teletype communications. Alternate routing is normally available, but the destruction of any portion of the system tends to overload the remaining facilities.

(1) Long distance telephone and teletype communications use cable or a combination of cable and radio relay. The cable may be aerial or underground. In populated areas the cable normally follows the roads, whereas in unpopulated areas it may run cross country. Underground cable usually runs cross country, but the route is marked for the convenience of the maintenance crews.

(2) Aerial cable can be destroyed by cutting the poles and cable. Underground cable often runs through concrete conduits and requires more time to destroy. Repair of cable can be delayed by removing a section of the cable. Destruction of telephone central offices and repeater stations causes greater damage and takes longer to repair than cutting the cables.

b. Radio stations may be located in well-protected areas and difficult to attack. However, antenna sites are often located a considerable distance away from the receiver and transmitter. These facilities are interconnected by transmission lines. Destruction of the antenna site and/or the transmission lines is usually easier to accomplish than destruction of the receiver or transmitter station.

124. Power Systems

Power lines are vulnerable to attack much in the same manner as wire communications. Large transmission towers often require demolitions for destruction. Critical points in any power system are the transformer stations. If these stations are not accessible to attacks by guerrilla units, long-range fire from small or large caliber weapons may disrupt their operations. Power producing plants and steam generating plants may be too heavily guarded for raid operations. To disable them, UW forces should concentrate on cutting off the fuel supply.

125. Water Supply Systems

The disruption of water lines supplying industries can often be profitably accomplished; water supplies generally are conducted through underground pipe lines, and may be destroyed with explosive charges. Raids against reservoir facilities and purification plants also are feasible, but the possible effects upon the civilian population must be considered.

126. Fuel Supply Systems

Petroleum and natural gases for an industrial area usually are supplied by pipe lines; damage to lines inflicted by rupture and ignition of fuel is considerably greater than damage inflicted on water lines. Large storage tanks at either end of a pipe line are highly vulnerable to weapons fire, especially when using incendiary projectiles. Contaminating agents may be injected into pipe lines or fuel tanks.

Section IV. DEFENSIVE OPERATIONS

127. General

Guerrilla operations are primarily offensive in nature. Guerrilla units with their relatively light weapons and equipment are normally inferior in strength and firepower to organized enemy forces. They should not, therefore, undertake defensive operations unless forced to do so or in support of special operations conducted by other theater forces. When the enemy attacks, guerrillas defend themselves by movement and dispersion, by withdrawals, or by creating diversions. Whenever possible, defensive operations are accompanied by offensive actions against the enemy's flanks and rear.

128. Preparation Against Enemy Offensives

a. Adequate intelligence measures normally provide advance warning of impending large-scale counter guerrilla operations. Guerrilla commanders must be cognizant of the following activities or conditions which might indicate impending enemy offensives in their operational areas

(1) Advent of suitable weather for extensive field operations.

(2) Arrival of new enemy commanders.

(3) Any change in the conventional battle situation which releases additional troops for counter guerrilla operations. Such changes include enemy victories over allied conventional forces, a lull in active operations, and a reduction of the size of the battle area.

(4) Increase in the size of local garrisons or the arrival of new units in the area, especially if these are combat troops or troops with special counter guerrilla capabilities such as radio direction finding units, CBR units, rotary winged aircraft, mountain, airborne, or reconnaissance troops.

(5) Extension of enemy outposts, increased patrolling and aerial reconnaissance.

(6) Increased enemy intelligence effort against the guerrillas.

b. Upon receiving information that indicates the enemy is planning a counter guerrilla campaign, the commander should increase his own intelligence effort, determine the disposition and preparedness of his subordinate units and review plans to meet the anticipated enemy action.

129. Defensive Measures

a. To divert the enemy's attention the commander directs that diversionary activities be initiated in other areas. Likewise he may intensify his operations against enemy lines of communications and installations. Full utilization of underground and auxiliary capabilities assists diversionary measures.

b. In preparing to meet enemy offensive action, key installations within a guerrilla base are moved to an alternate base and essential records and" supplies are transferred to new locations while those less essential are destroyed or cached in dispersed locations. In the event that the commander receives positive intelligence about the enemy's plans for a major counter guerrilla operation, he may decide to evacuate his bases without delay.

c. The commander may decide to delay and harass the advancing enemy. Here his object is to make the attack so expensive that the enemy will terminate operations and be, content with his original dispositions. First, security activities on the periphery as well as within a, base are accelerated. Maximum utilization is made of the defensive characteristics of the terrain; ambushes are positioned to inflict maximum casualties and delay; antipersonnel mines are employed extensively to harass the enemy. As the enemy overruns various. strong points, the defenders withdraw to successive defensive positions to delay and harass again. When the situation permits, they may disperse, pass through the line of encirclement, and initiate attacks on the enemy's flanks, rear and supply lines. If the enemy is determined to continue his offensive, the guerrilla forces should disengage and evacuate the area. Under no circumstances does the guerrilla force allow itself to become so engaged that it loses its freedom of action and permits enemy forces to encircle and destroy it.

d. When faced with an enemy offensive of overwhelming strength, the commander may decide to disperse his force, either in small units or as individuals to avoid destruction. This course of action should not be taken unless absolutely necessary inasmuch as it makes the guerrilla organization ineffective for a considerable period of time.

130. Encirclement

An encircling maneuver is the greatest danger to guerrilla forces because it prevents them from maneuvering. Once the enemy has succeeded in encircling a guerrilla force, he may adopt one of several possible courses of action (fig. 35) .

a. The simplest is to have his troops close in from all sides, forcing the guerrillas back until they are trapped in a small area which is then assaulted. Differences in terrain make it almost impossible for his troops to advance at an equal rate all around the perimeter, thus creating the possibility of gaps between individuals and units.

b. In other cases the enemy may decide to break down the original circle into a number of pockets which will be cleared one by one. The creation of these pockets 16 a repetition of the Original encirclement. In this situation the guerrillas must either break out or escape through gaps, which may appear as enemy forces are maneuvering into new positions.

c. Perhaps the most difficult situation for guerrillas to counter with is an assault after encirclement has been accomplished. In this maneuver enemy forces on one side of the encircled area either dig in or use natural obstacles to block all possible escape routes, while the forces on the opposite side advance driving the guerrillas against the fixed positions. As the advance continues, enemy forces which were on the remaining two sides are formed into mobile reserves to deal with any breakouts (fig. 36).

131. Defense Against Encirclement

a. Initial Actions. A guerrilla commander must be constantly on the alert for indications of an encirclement. When he receives indications that an encircling movement is in progress such as the appearance of enemy forces from two or three directions, the guerrilla commander immediately maneuvers his forces to escape while enemy lines are still thin and spread out, and coordination between advancing units is not yet well established. Records and surplus equipment are either cached or destroyed. Thus, the guerrilla force either escapes the encirclement or places itself in a more favorable position to meet it. If for some reason, escape is not initially accomplished, movement to a ridge line is recommended. The ridge line affords observation, commanding ground, and allows movement in several directions. The guerrillas wait on this high ground until periods of low visibility or other favorable opportunity for a break-through attempt occurs.

b. Breakout (fig. 37). Two strong combat detachments precede the main body which is covered by flank and rear guards. If gaps between the enemy units exist, the combat detachments seize and hold the flanks of the escape route. When there are no gaps in the enemy lines, these detachments attack to create and protect an escape channel. The break-through is timed to occur during periods of poor visibility, free from enemy observation and accurate fire. During the attempt, guerrilla units not included in the enemy circle make attacks against his rear to lure forces away from the main break-out attempt and help to create gaps. After a successful break-through, the guerrilla force should increase the tempo of its operations whenever possible, thus raising guerrilla morale and making the enemy cautious in the future about leaving his bases to attack the guerrilla areas.

c. Action If Breakout Fails (fig. 38). If the breakout attempt is unsuccessful, the commander divides his force into small groups and instructs them to infiltrate through the enemy lines at night or hide in the area until the enemy leaves. This action should be taken only as a last resort, as it means the force will be inoperative for a period of time and the morale of the unit may be adversely affected. Reassembly instructions are announced before the groups disperse.

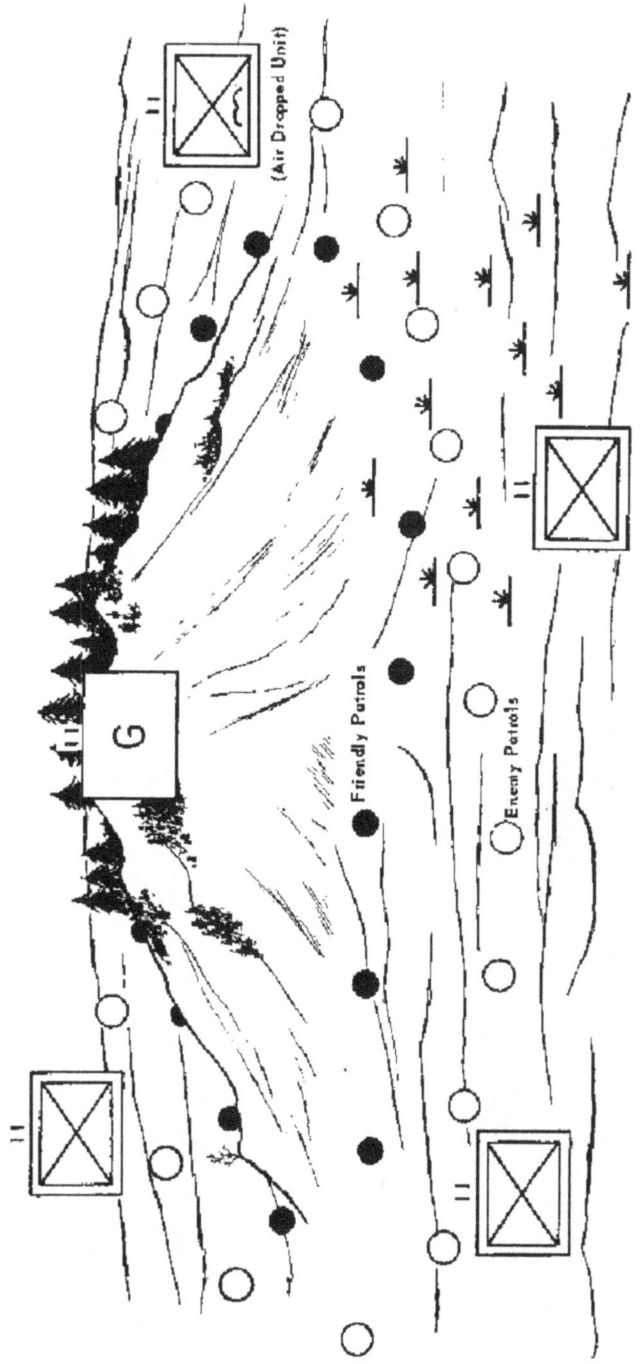

Figure 35. Encirclement.

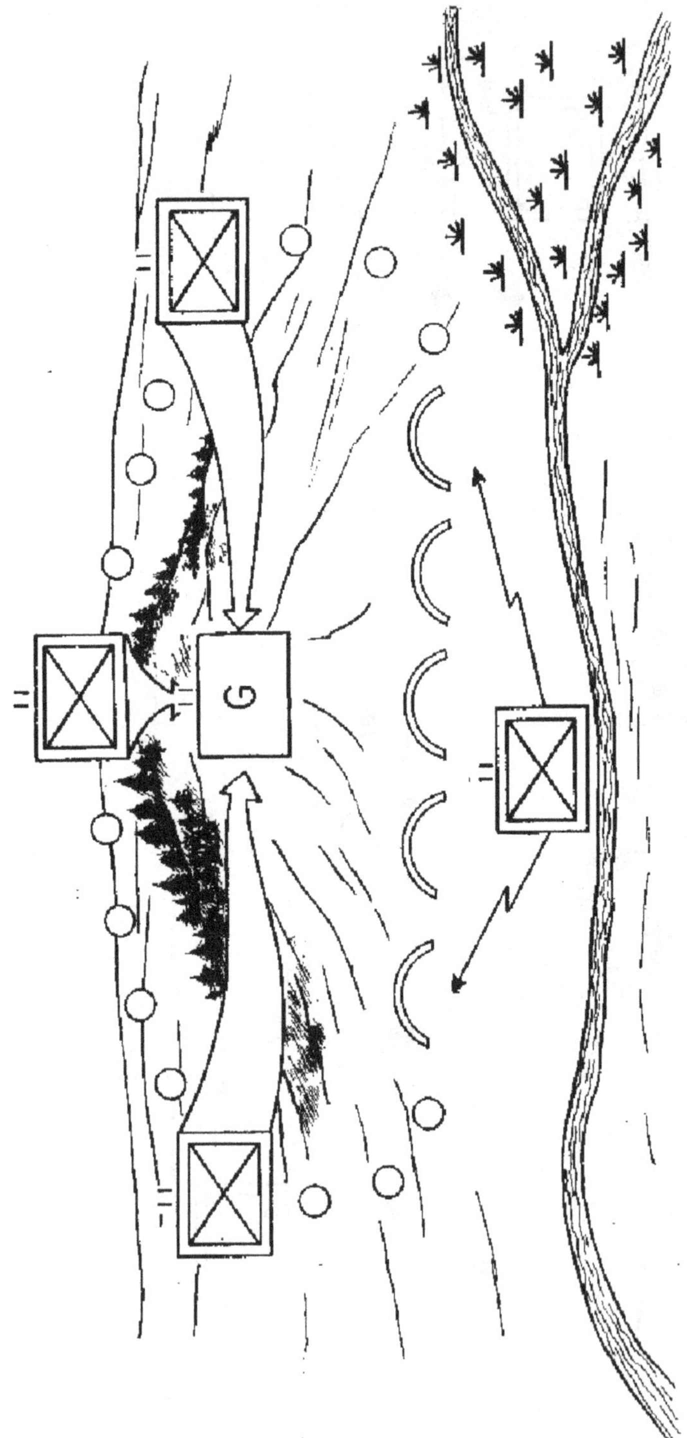

Figure 36. Encirclement and assault.

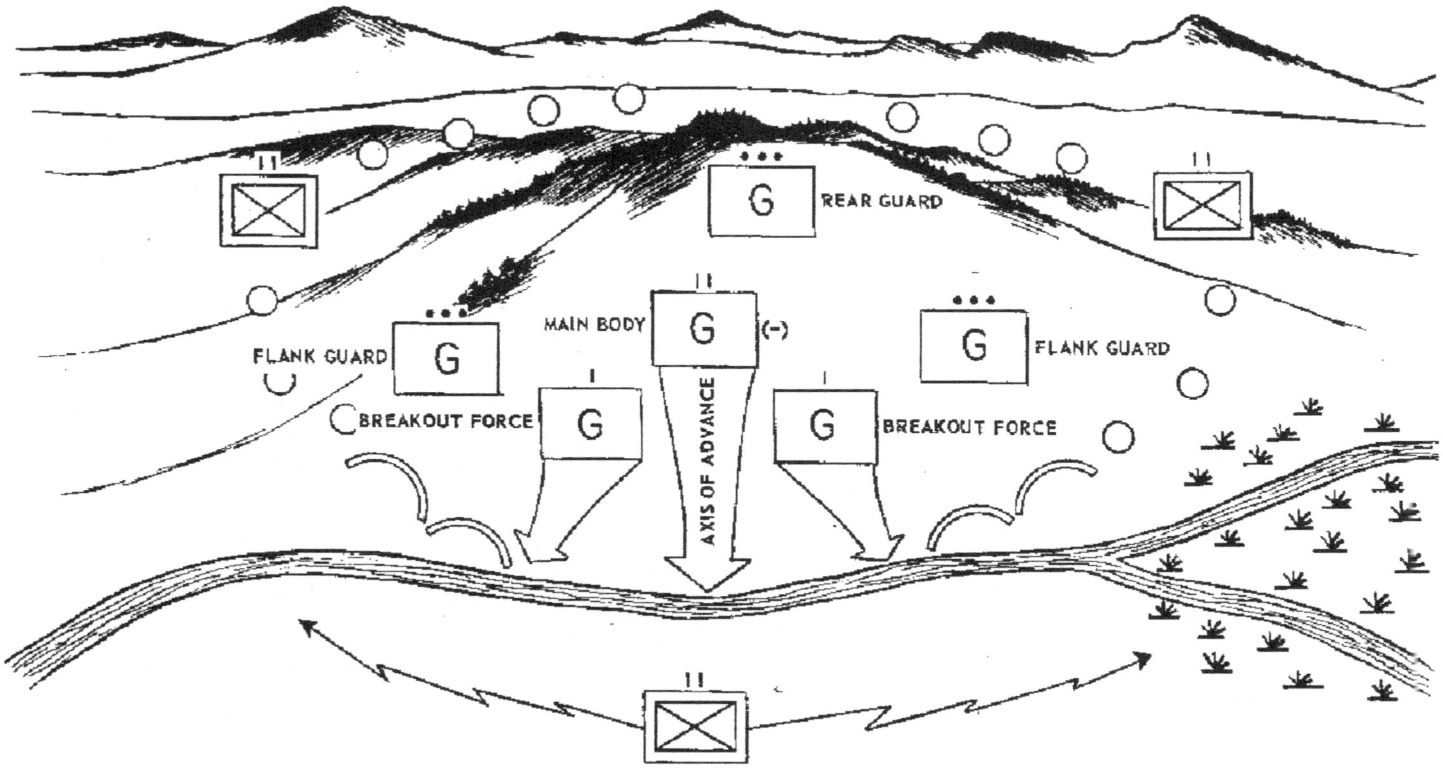

Figure 37. Guerrilla breakout from encirclement.

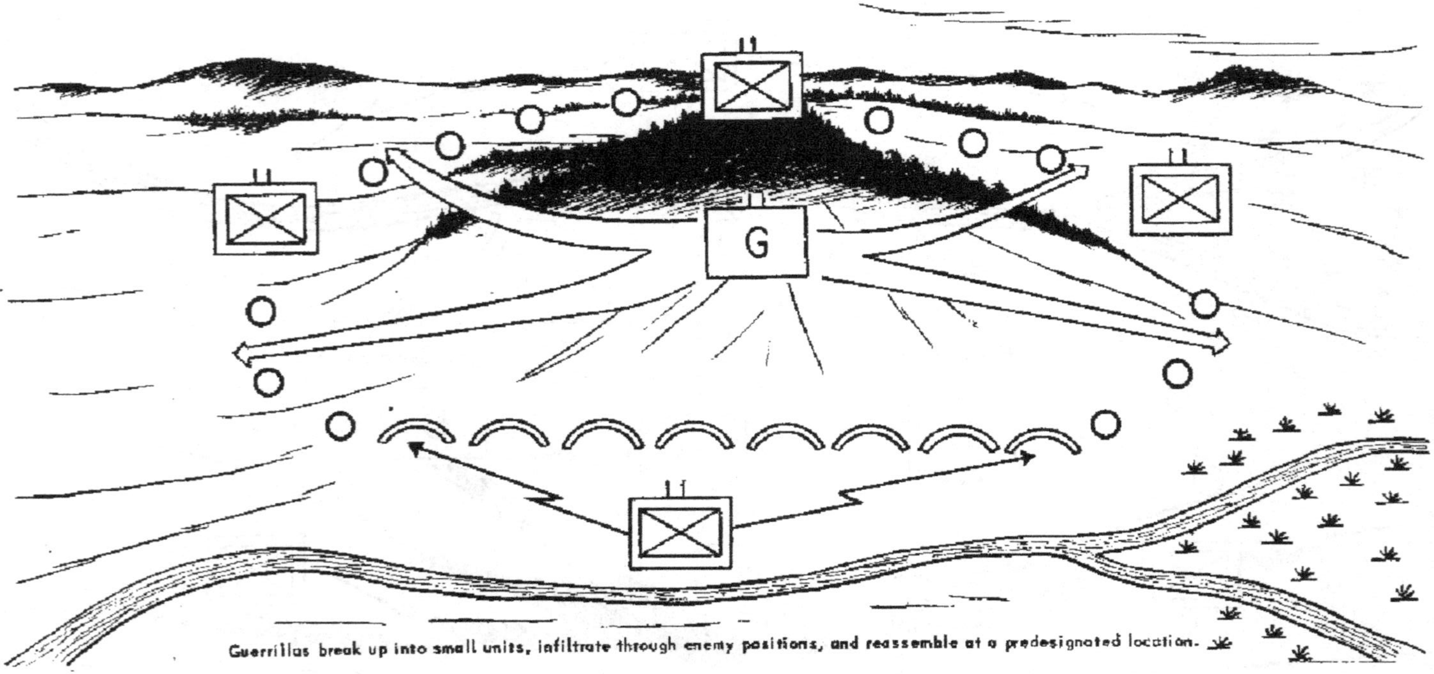

Figure 38. Action if breakout fails.

Section V. EMPLOYMENT OF UNCONVENTIONAL WARFARE FORCES TO ASSIST CONVENTIONAL FORCES' COMBAT OPERATIONS

132. General

When the area of influence of the field army (or other conventional force command) overlaps a guerrilla warfare area, operational control of the guerrilla forces concerned is passed to the field army (or other conventional force command) commander. Interdiction operations are of greater immediate importance and are planned to support tactical objectives. Attacks against enemy supply depots, lines of communications and other installations essential to support of his tactical troops increase. The psychological impact of the success of friendly conventional forces is magnified by intensified UW activity. Psychological warfare efforts exploiting these conditions are expanded. Enemy tactical targets are located and reported to conventional forces on an ever-mounting scale, thus supporting the increased range of modern weapons. Evasion and escape operations expand to handle large numbers of friendly personnel who may find themselves evaders. In addition to the aforementioned tasks, guerrilla forces can expect missions which directly assist combat operations of friendly tactical units. Although primarily of value in support of the tactical offense, guerrilla warfare can also assist friendly forces engaged in defensive operations. During the period of operations to assist conventional forces, link-up between friendly tactical commands and guerrilla forces usually takes place.

133. Missions

In addition to an acceleration of activity discussed in paragraph 132, guerrilla forces can assist the combat operations of conventional forces engaged in envelopment or penetration operations. Examples of missions appropriate for guerrilla forces to assist field army (or other conventional force commands) are

a. Seizure of hey terrain to facilitate airborne an u amphibious operations. This may include portions of the airhead or beachhead line, drop and landing zones or reconnaissance and security positions.

b. Employment as a reconnaissance and security force.

c. Seizure of key installations to prevent destruction by the enemy. Examples are bridges, defiles, tunnels, darns, etc.

d. Diversionary attacks to support friendly cover and deception operations.

e. Operations which isolate selected portions of the battle area, airborne objective area or beachhead.

134. Special Considerations

a. Tactical commanders who employ guerrilla forces must carefully consider their capabilities when assigning them operational tasks. Guerrilla units are organized and trained to execute planned offensive operations in enemy controlled areas. The sustained combat capabilities of guerrilla units are affected by several variables such as: size, organization, leadership, training, equipment, background of personnel and extent of civilian support. These differences are usually more pronounced among guerrilla units than in conventional organizations of comparable strengths. Consequently, like-size guerrilla units may not be capable of accomplishing comparable missions. Assignment of missions to guerrilla units should take advantage of their light infantry characteristics and area knowledge. Attached Special Forces liaison personnel recommend to the tactical commander appropriate tasks for guerrilla forces.

b. Perhaps the severest limitation common to guerrilla forces when employed with friendly tactical units is their shortage of adequate voice communications equipment and transportation. This is particularly true when guerrilla units are operating with a mobile force in a penetration, envelopment or exploitation. For this reason guerrilla units have a slower reaction time in terrain favoring a high degree of mechanical mobility. Conventional commanders may overcome this disadvantage by providing the necessary equipment or utilizing the guerrilla force on an area basis. For further discussion, see paragraphs 136 through 138.

c. Another special consideration is the requirement for restrictions in the use of nuclear weapons and CBR agents by other friendly forces. This is particularly true when a large segment of the civilian population supports the resistance movement. Careful coordination of targets selected for nuclear and CBR attack is required between the conventional force commander and the guerrilla force. Provisions must be made to provide adequate warning to friendly elements of the population who may be /endangered by nuclear and CBR weapons.

135. Command Relationships

a. General (figs. 39 and 40). When operational control of the UW forces is passed to the field army (or other conventional force command) commander, administrative support of the guerrilla warfare area is retained by ,the SFOB. Concurrent with the change in

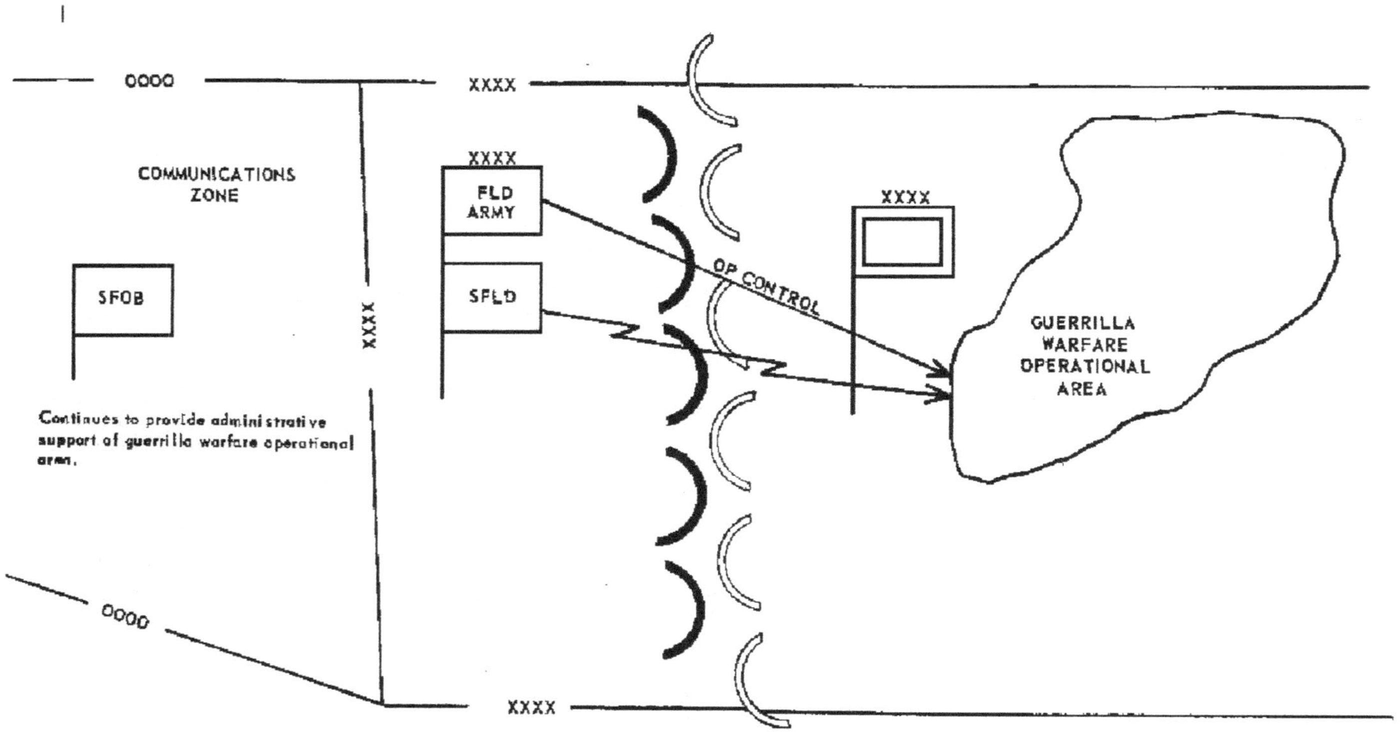

Figure 39. Operational control exercised by a conventional command.

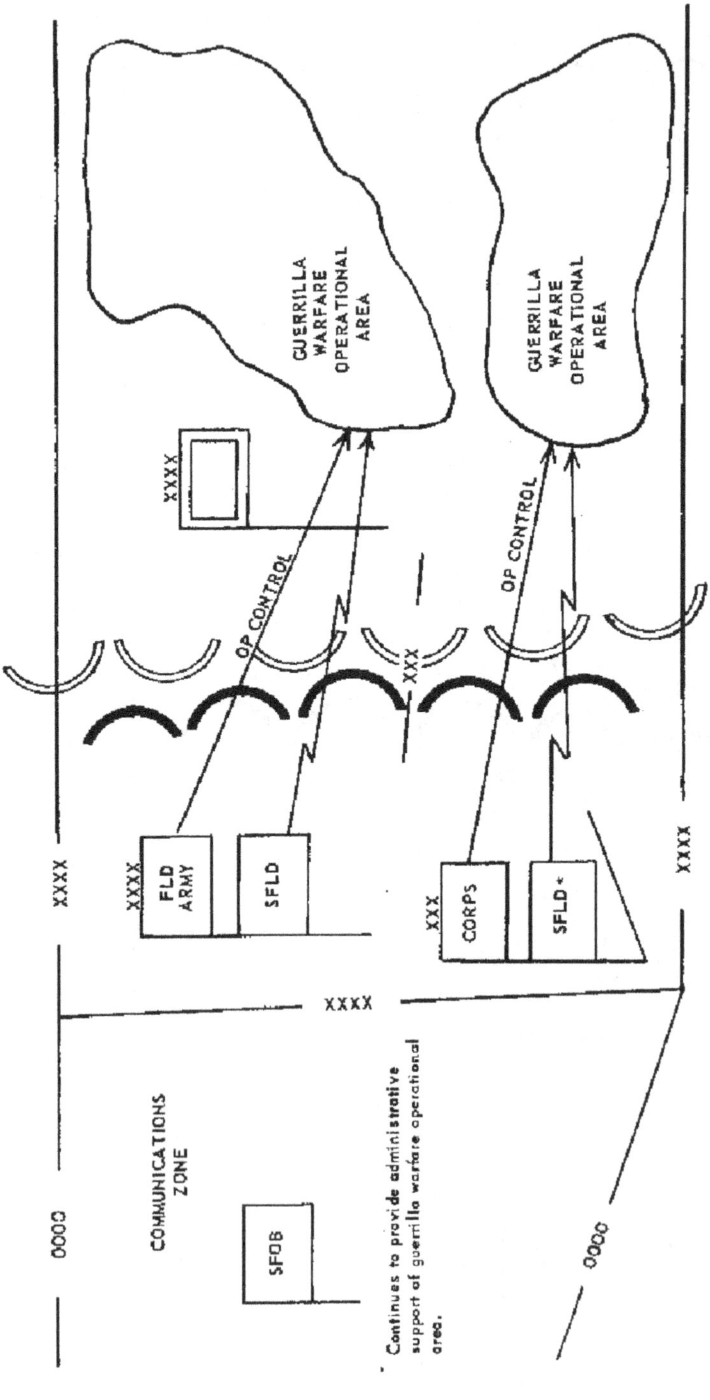

*Personnel drawn from field army SFLD.

Figure 40. Operational control exercised by multiple conventional commands.

operational control of the guerrilla force from theater to tactical command level, the Special Forces group commander provides a liaison detachment to the headquarters of the command concerned. The Special Forces liaison detachment assists the tactical commander in the direction and coordination of attached guerrilla forces. Operational control of part of all of the guerrilla force may be passed to lower tactical echelons as required but is normally not delegated below division level.

b. Special Forces Liaison Detachment.

(1) *Composition.* The Special Forces liaison detachment is a non-TOE team which may vary from a minimum of one liaison officer to a modified operational detachment C or B. The size and composition of the liaison detachment is dictated by the type headquarters having operational control; size, command structure, and disposition of guerrilla forces concerned; and availability of required communication equipment.

(2) *Functions.* The liaison detachment assists the tactical commander in the coordination of Special Forces directed administrative operations and tactical unit directed UW operations. The detachment commander

(a) Plans and recommends employment of guerrilla forces.

(b) Exercises operational control over guerrilla forces when this authority is delegated by the tactical commander.

(c) Maintains liaison with subordinate tactical head quarters as directed.

(d) Maintains liaison with Special Forces group commander.

c. Communications. Communications between the SF liaison detachment and operational areas may be established in several ways

(1) The liaison detachment may have a direct link to the operational area (1, fig. 41) . In this situation, additional radio equipment is provided by the SFOB for the liaison detachment base station. The advantage is direct communications. The disadvantage is that additional equipment and personnel usually must be provided by other theater signal sources.

(2) The SFOB may act as the radio intermediary between the liaison detachment and the operational area (2, fig. 41).

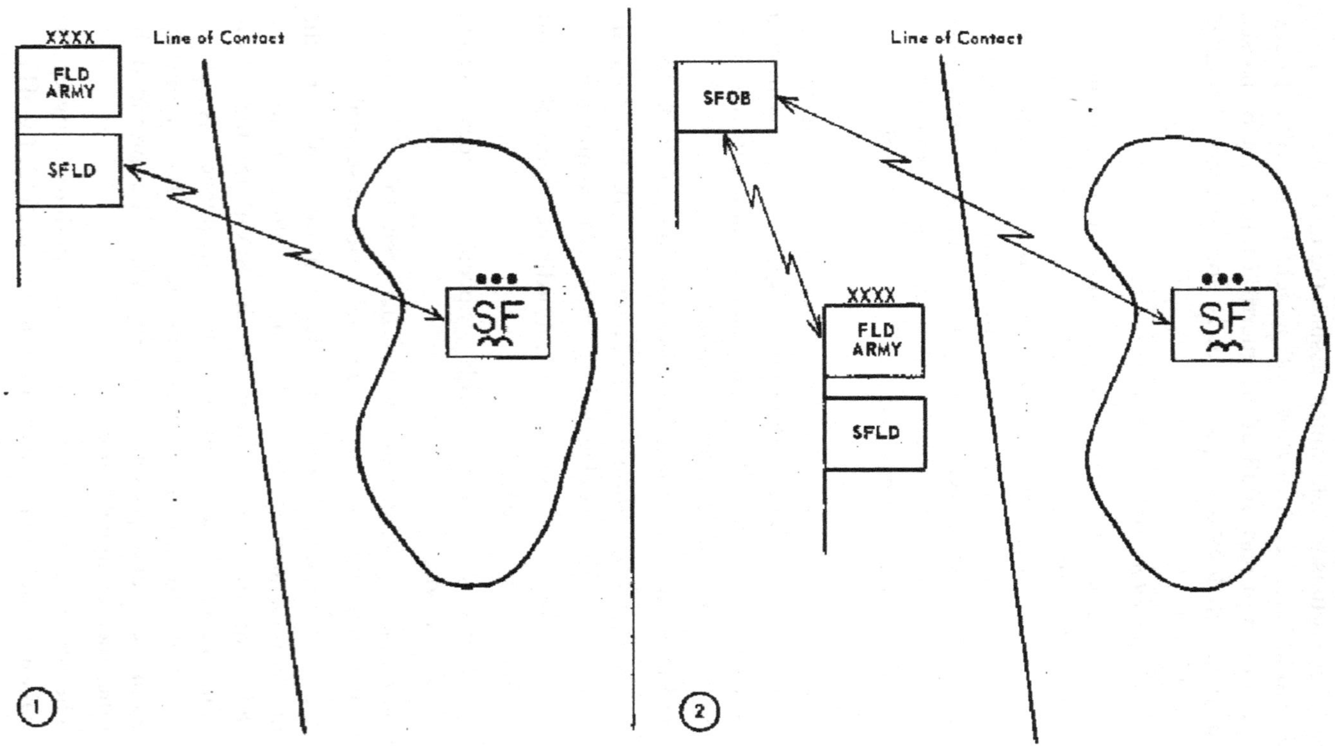

Figure 41. Communications between the conventional command and guerrilla warfare operational areas.

In this situation messages are relayed from the tactical command headquarters via the SFOD to the operational detachment. Communications from the detachment utilize the reverse sequence. This system has the advantage of utilizing established communication facilities and requires no additional communication equipment and personnel with the SF liaison detachment. However, the time lapse between initiation and receipt of messages is increased.

(3) A variation of the solution cited in c(2) above may be adopted when both senior and subordinate tactical commands control different elements of the guerrilla force. For example, both field army and corps control guerrilla forces, yet insufficient communications equipment is available to provide both headquarters with a base station. The SF liaison detachment locates its base station at field army headquarters and corps relays instructions to guerrilla units under its control via the Special Forces radio facility at army. Special forces liaison detachment personnel are located at both head,/quarters.

136. Support of Ground Offensive Operations

a. General. As the conventional force command's area of influence overlaps the guerrilla warfare operational area, guerrilla units shift to operations planned to produce immediate effects on enemy combat forces. Initially, these activities are directed against the enemy communication zone and army support troops and installations. As the distance between guerrilla and conventional forces decreases, guerrilla attacks have greater influence on the enemy combat capability. Guerrilla operations support penetrations and envelopments and are particularly effective during exploitation and pursuit

b. Guerrilla Operations During a Penetration. Due to the high density of enemy combat troops in the immediate battle area, guerrillas can give little direct assistance to friendly forces in initial phases of a penetration (rupture of the enemy defensive position or widening the gap). Guerrilla forces can best support the attack by isolating, or assisting in the seizure, of the decisive objective (fig. 42). Guerrilla forces hinder or prevent movement of enemy reserves, interrupt supply of combat elements, and attack his command and communications facilities, fire support means and air fields. Locations of critical installations and units which the guerrillas cannot effectively deal with are reported to the tactical commander for attack. As friendly forces near the decisive objective, guerrilla units direct their operations toward isolating the objective from enemy reserves. In some instances guerrilla forces may be able to seize and hold the objective or key approaches to it for a limited time pending link-up with the conventional force.

c. Guerrilla Operations During An Envelopment.

(1) Guerrilla units assist the enveloping force in much the same way as in a penetration (fig. 43). Guerrillas can conduct diversionary attacks to assist other forces' cover and deception plans. As in the penetration, guerrillas hinder movement of reserves, disrupt supply, attack command and communications installations and reduce the effectiveness of enemy fire support. They may assist in containment of. bypassed enemy units. They attempt to isolate the objective of the enveloping force. They may seize and hold critical terrain, such as bridges, defiles and tunnels, to prevent enemy destruction. They may perform screening missions to the front and flanks or be a security element to fill gaps between dispersed units of the enveloping force.

(2) If used in a reconnaissance or security role, guerrilla units operate on an area basis. That is, they perform their security or screening role within a specified area during the time the enveloping force passes through the area. Guerrilla units usually do not possess the transportation or communications to accompany mobile forces.

d. Guerrilla Operations During Exploitation. As friendly tactical units pass from a successful penetration or envelopment to the exploitation of their gains, guerrilla operations increase in effectiveness. As the enemy attempts to reconstitute an organized defense or withdraw to new positions he is attacked at every opportunity by UW forces (fig. 44). Enemy troops, normally available for rear area security duties, are committed to attempts to restore his defensive position, thus enabling guerrilla attacks to be increased in scope and magnitude against rear area installations whose capability for defense is reduced. Guerrilla forces assist in containing bypassed enemy units, rounding up stragglers and prisoners, seizing control of areas not occupied by the exploiting force, attacking enemy units and installations and adding to the general demoralization caused by the exploitation and subsequent pursuit. As link-up with the exploiting force is accomplished, guerrilla forces may be employed as discussed in paragraphs 140 through 147.

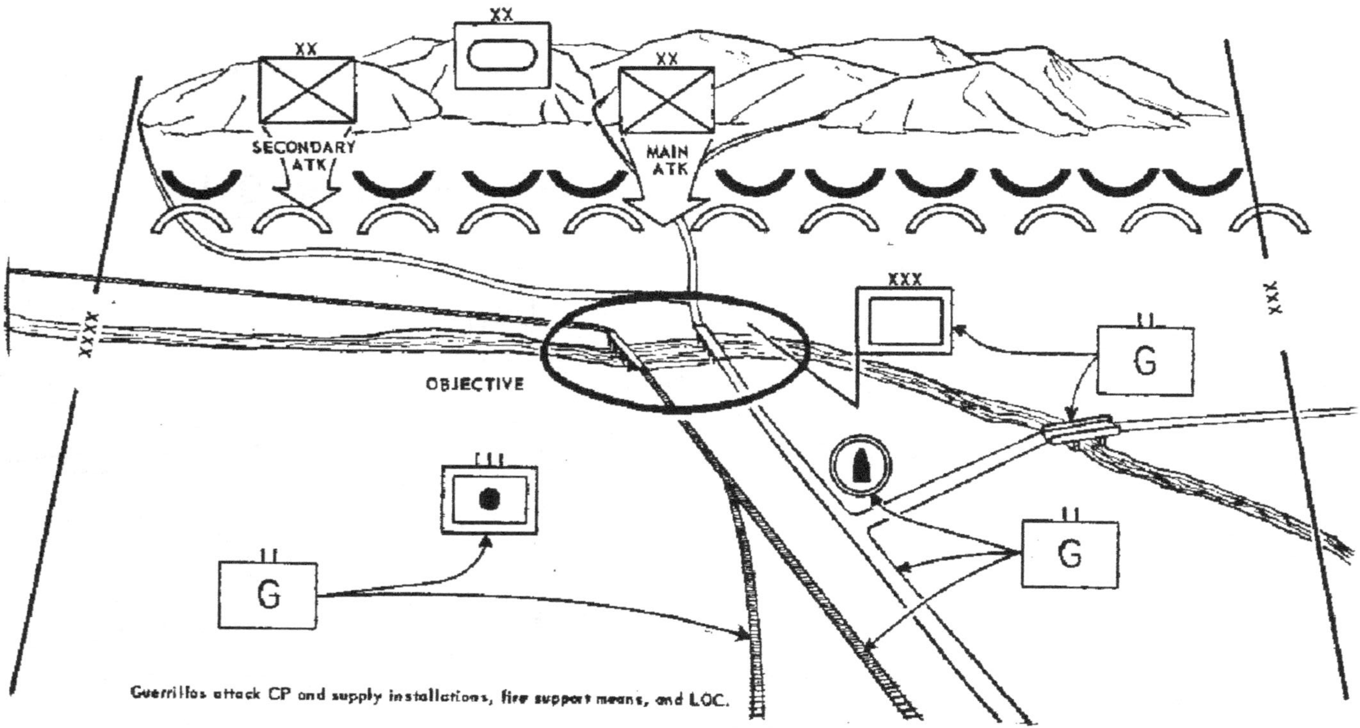

Figure 42. Guerrilla operations to assist a penetration.

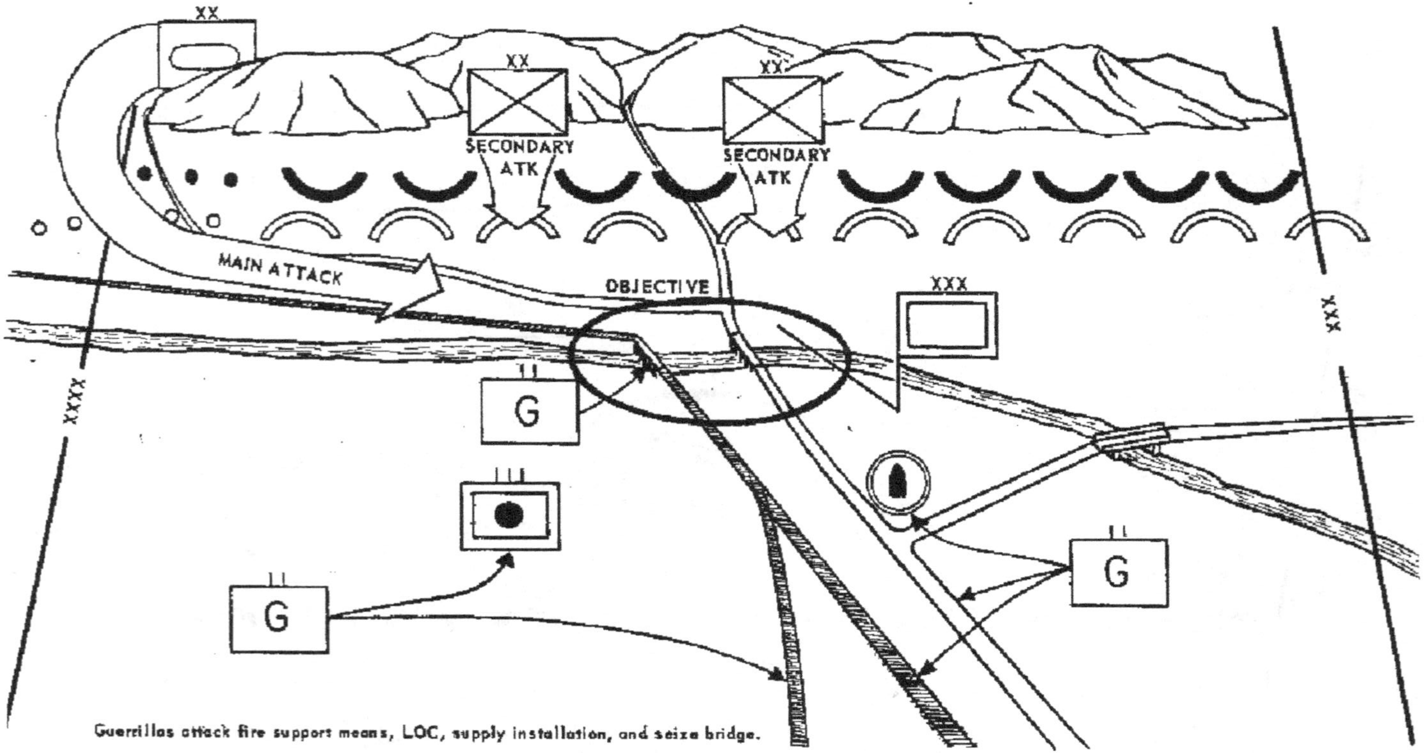

Guerrillas attack fire support means, LOC, supply installation, and seize bridge.

Figure 43. Guerrilla operations to assist in envelopment.

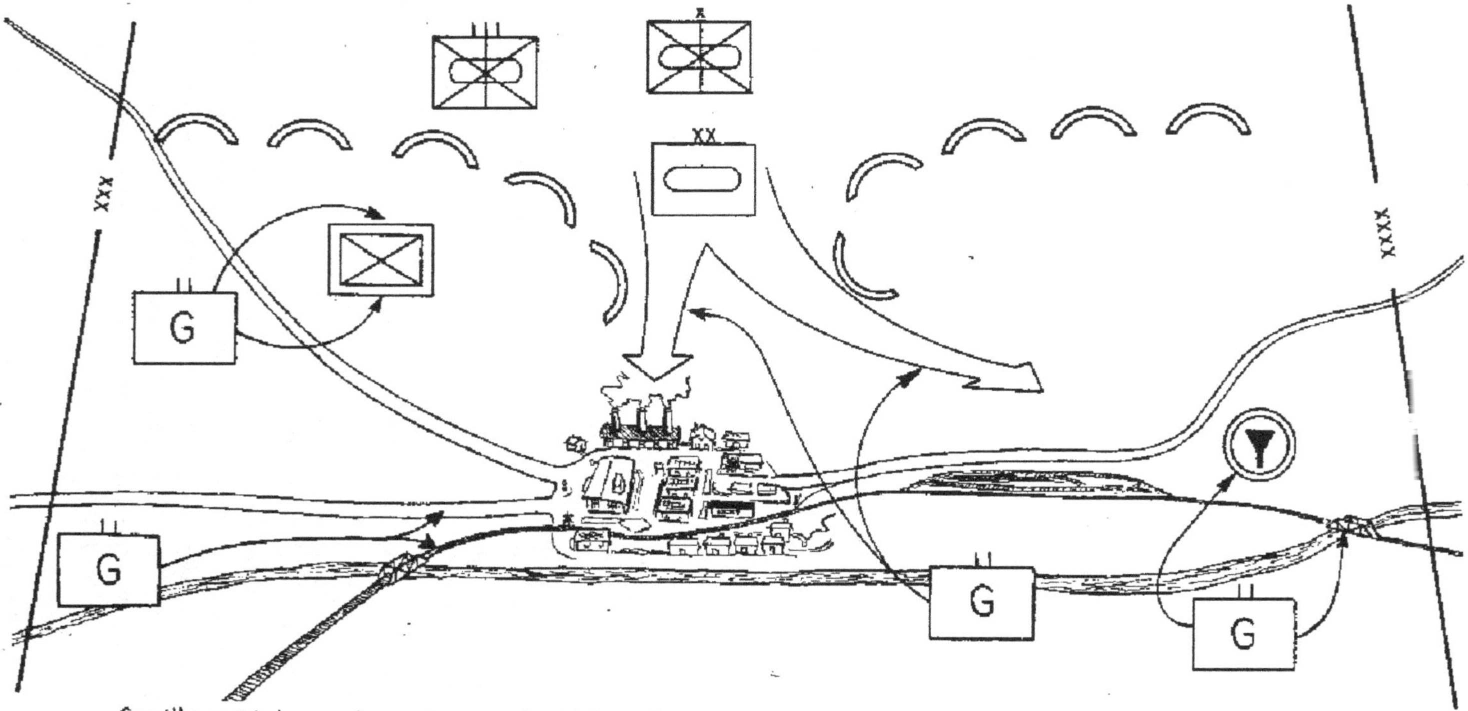

Guerrillas contain by-passed enemy force, interdict LOC to isolate objective area, screen gap between attacking units, attack supply installations and seize bridges to prevent destruction by enemy.

Figure 44. Guerrilla operations during exploitation.

e. Command Relationships. Operational control of the guerrilla force is retained at the level best able to coordinate the actions of the operation. As link-up becomes imminent guerrilla units nearest the attacking force may be attached to or placed under the operational control of that force. Concurrent with link-up, responsibility for administrative support of the guerrilla force is passed to the tactical command. When link-up has been effected the utilization of guerrilla forces is in consonance with guidance provided by the theater commander. See paragraphs 140 through 147 for post link-up employment.

137. Support of Airborne Operations

a. General.

(1) Guerrilla forces, by virtue of their location in enemy controlled areas, can materially assist conventional forces engaged in airborne operations. They support airborne forces during the assault phase and subsequent opera dons. They may also be employed in conjunction with airborne raids and area interdiction operations.

(2) For details of link-up between airborne and guerrilla forces, see paragraph 139.

b. Guerrilla Assistance to an Airborne Assault (fig. 45).

(1) Initially, UW forces can provide selected current intelligence of the objective area upon which the airborne force commander bases his plans. Immediately prior to the assault, guerrilla units may be able to secure drop and landing zones; seize objectives within the airhead line; and occupy reconnaissance and security positions. Concurrent with landing of the assault echelon, guerrillas can conduct reconnaissance and security missions; provide guides and information; interdict approaches into the objective area; control areas between separate airheads and dispersed units; attack enemy reserve units and installations; and conduct diversionary attacks as a part of the cover and deception plan. Additionally, UW forces may control civilians within the objective area.

(2) Correct timing of guerrilla operations with the airborne assault is essential. If committed prematurely, guerrilla forces may nullify the surprise effect of the operation and, in turn, be destroyed by the enemy. Conversely, if committed too late, the desired effects of the guerrilla force employment may never be realized.

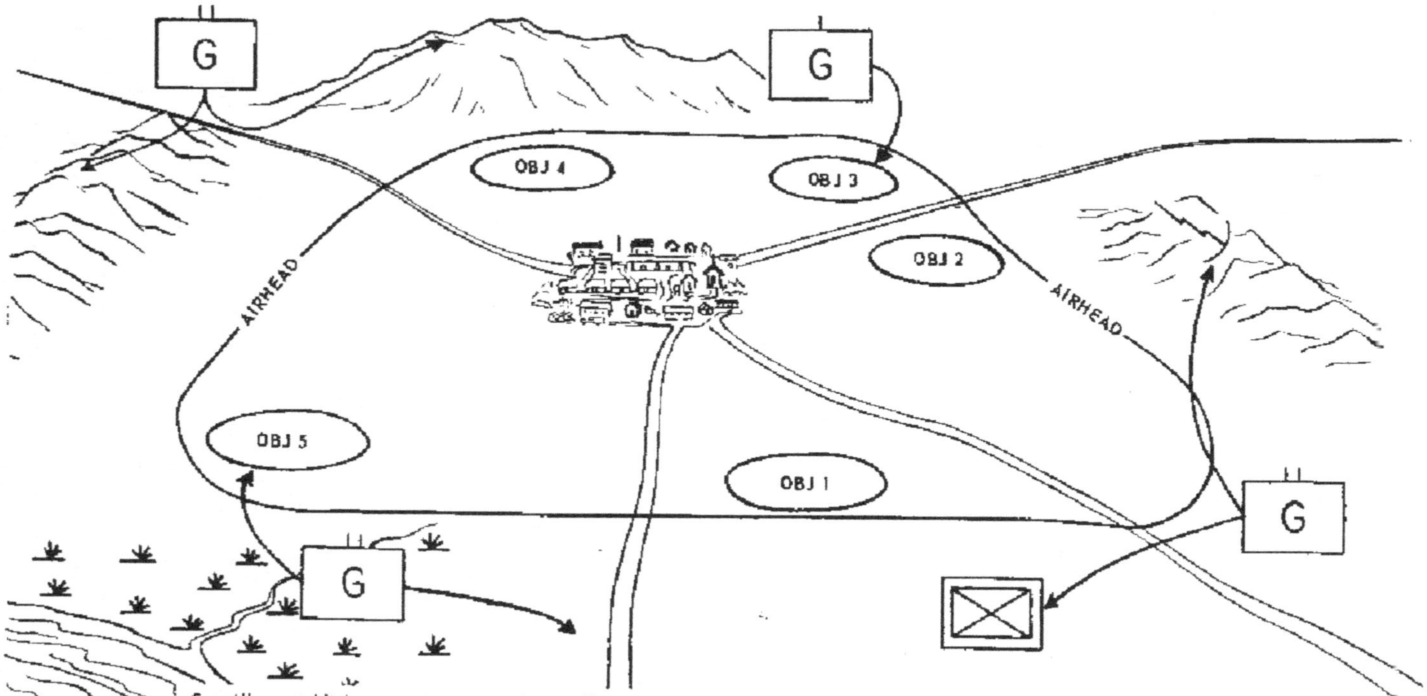

Guerrillas establish reconnaissance and surveillance positions, seize objectives, attack enemy units and radar installation.

Figure 45. Guerrilla operations to assist an airborne assault.

c. Guerrilla Assistance to Subsequent Operations. As the assault phase of an airborne operation passes into the defensive or offensive phase, UW forces continue to exert pressure on the enemy forces in the vicinity of the objective area. Guerrillas continue to provide up-to-date information on enemy moves and disposition. Attacks are directed against enemy units attempting to contain or destroy the airborne force, thus requiring him to fight in more than one direction. Airborne forces which have an exploitation mission may employ recovered guerrilla units in reconnaissance and security roles as guides and to assist in control of void areas between dispersed units. If the airborne force is to be withdrawn, the guerrillas can assist to cover the withdrawal by diversionary operations conducted in the rear of enemy forces.

d. Airborne Raids. Guerrilla forces assist airborne raids in a similar fashion as they do the assault phase of an airborne operation. They provide information and guides; perform reconnaissance and security missions and divert enemy forces during the withdrawal of the raiding force. An additional factor to consider before using guerrilla forces to support an airborne raid is the undesirable effect of enemy reaction on resistance organizations after withdrawal of the raiding force.

e. Area Interdiction Operations. Airborne units are seldom committed to guerrilla warfare areas to conduct interdiction operations if the guerrilla force has the capability to conduct such operations. However, in areas where no effective resistance exists, airborne forces may be committed to conduct interdiction operations. Whatever guerrilla forces are located in areas selected for airborne interdiction, assist the airborne force to conduct their operations. They provide intelligence information and guides; conduct reconnaissance and security missions; control the civilian population; assist in collecting supplies and generally aid the airborne force commander in making the transition from conventional operations to. guerrilla operations. Special forces detachments, if available, may conduct special training within the operational area to increase the capability of the airborne force in guerrilla warfare techniques.

f. Command Relationships.

(1) Operational control of guerrilla forces within the objective area or influencing the mission of the airborne force is exercised by the airborne force commander. Control of other guerrilla forces whose effect upon the airborne operation is indirect is initially retained by the commander directing the

airborne operation (joint airborne task force or theater army commander).

(2) Concurrent with link-up, responsibility for administrative support of the guerrilla force is passed to the link-up force. For employment of guerrilla forces after link-up, see paragraphs 140 through 147.

138. Support of Amphibious Operations

(fig. 46)

a. General.

(1) Guerrillas support conventional forces engaged in amphibious operations, generally in one or more of the following ways (fig. 22)

(a) By conducting operations to hinder or deny the enemy approach to the beachhead.

(b) By seizing and holding all or a portion of the beach head.

(c) By assisting airborne operations which are a part of or complement the amphibious assault.

(d) By conducting cover and deception operations to deceive the enemy as to the location of the actual beachhead.

(2) Guerrilla forces operating within the objective area will be assigned to the operational control of the amphibious task force commander when he becomes responsible for the objective area. Operational control of guerrilla forces is further assigned to the landing force commander when he assumes responsibility for operation ashore. Normally, operational control of guerrilla forces assisting amphibious operations is not passed below divisional level. Concurrent with link-up, responsibility for administrative support of the guerrilla force is passed to the link-up force.

b. Guerrilla Assistance to an Amphibious Assault.

(1) If the selected beachhead is defended in strength, guerrilla operations are planned to hinder or deny the enemy approaches into the beachhead area. By prearranged plan, guerrilla units interdict approaches into the area; attack reserves; destroy command and communications facilities; logistical installations and airfields which can support the enemy defense forces in or near the beachhead. Fire support elements within range of the beachhead are a primary guerrilla target.

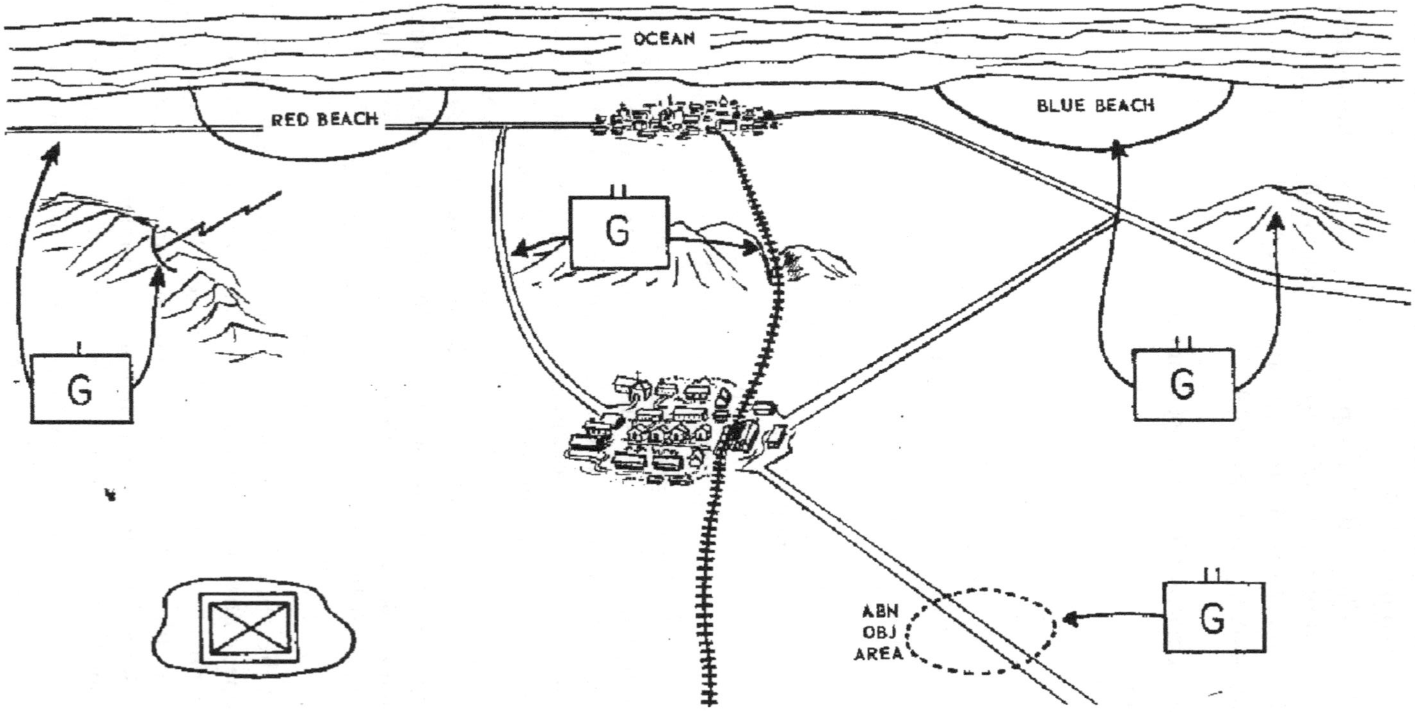

Guerrillas interdict beach approaches, attack radar installation, seize portion of beach head line, assist airborne assault.

Figure 46. Guerrilla operations to assist an amphibious landing.

(2) If the selected beachhead is lightly defended or undefended, guerrilla units may seize and hold portions of the beachhead. Guerrilla farces seize their objectives just prior to the initial assault. When required, landing force unit tasks must provide for early relief of guerrilla units. Plans for naval fire support to guerrilla forces must include provisions for the conduct and adjustment of fires. Naval liaison personnel, shore fire control parties, and tactical air control parties will be attached when required. The size of the beachhead, enemy situation and size of the guerrilla force govern the extent of the beachhead to be allotted to the guerrillas. For employment of guerrilla forces after link-up, see paragraphs 140 through 147.

(3) If are airborne operation is conducted as a part of or to complement the amphibious operation, guerrillas can be employed as described in paragraph 137.

(4) Guerrillas may be employed in a cover and deception role to assist amphibious assaults. Guerrilla forces intensify operations in selected areas to deceive the enemy as to the exact location of the main landings. Air defense radar and coastal detection stations are targets for guerrilla attack to reduce the enemy's early warning capability. Rumors as to time and place of landing may be spread among the population. A sudden increase in, or cessation of resistance activities tends to keep the enemy on edge and uncertain. The employment of the guerrilla force in support of cover and deception is integrated into the overall amphibious operation plan.

(5) Guerrilla operations in support of the landing force after completion of the assault phase and termination of the amphibious operation are as discussed in paragraphs 136 and 140 through 147.

(6) As in airborne operations, timing of the use of guerrilla forces in relation to the amphibious operation is extremely important, Premature commitment alerts the enemy and may lead to the destruction of the guerrilla force. Conversely, late employment may not have the desired effect upon the enemy.

139. Link-Up Operations

a. General.

(1) Most offensive operations in which guerrilla forces assist tactical commands involve a juncture between elements of

the two forces. Normally during link-up operations, the guerrilla force is the stationary force, and the conventional unit the link-up force.

(2) Not all guerrilla forces in an operational area are involved in link-up with tactical units. Some guerrilla units may be assigned missions assisting tactical commands where the requirements of the operation preclude physical juncture. For example, during a raid or area interdiction operations by airborne forces or when conducting operations as part of a cover and deception plan for an amphibious force, it is often undesirable to link-up all guerrilla units with the attacking units.

(3) Concurrent with link-up responsibility for administrative support of the guerrilla force passes from the SFOB to the link-up force.

(4) Regardless of the conditions under which link-up occurs, the following considerations govern planning:

(a) Command relationships.

(b) Liaison.

(c) Coordination of schemes of maneuver.

(d) Fire coordination measures.

(e) Communications coordination.

(f) Employment following link-up.

b. Command Relationships. Operational control of guerrilla forces is retained by the major link-up force until link-up is effected. For example, a division making an airborne assault exercises operational control of the guerrilla force. When link-up with guerrilla units is accomplished, these units may then be employed under division control or attached to subordinate elements such as a brigade on an independent or semi-independent mission. For a detailed discussion of command relationships in various situations, see paragraphs 136, 137, and 138.

c. Liaison.

(1) As operational control of guerrilla warfare areas are transferred from theater level to tactical commands, liaison personnel are exchanged between the SFOB and the tactical command concerned. The SFOB attaches a Special Forces liaison detachment to the tactical command headquarters. For composition and duties of this liaison detachment, see paragraph 135.

(2) As the distance between the tactical command and guerrilla forces decreases, operational control of the

guerrilla warfare area may be transferred to subordinate tactical elements. The SF liaison detachment furnishes necessary liaison personnel to these subordinate headquarters. In those instances where only one operational area exists the entire liaison detachment is attached to the subordinate headquarters.

(3) When link-up planning commences, provisions are made for an exchange of liaison personnel between the link-up force and the guerrilla warfare area command. A liaison party from the guerrilla force is exfiltrated. This party, consisting of SF and indigenous representatives, assists in the link-up planning for the tactical commander. The guerrilla force liaison personnel are able to provide the latest friendly and enemy situation and recommend link-up coordination measures and missions for guerrilla units.

(4) Shortly after removal of the guerrilla force's liaison party from the operational area, the tactical commander infiltrates his liaison party to join the area command. This liaison party consists of representatives from the G3 section, the Special Forces liaison detachment, tactical air control parties, forward observer tears, and communications personnel and equipment. The liaison party furnishes the guerrilla area commander the link-up plan and appropriate missions.

(5) Army aviation is generally used to transport liaison parties into and out of the operational area.

d. Coordination of Schemes of Maneuver. Standard control measures are established to assist link-up. See FM 57-30, FM 7-100, and FM 17-100 for details of these control measures. Guerrilla units are usually dispersed over a larger area, consequently link-up will take place at several widely separated areas, thus necessitating designation of more link-up points than normal.

e. Fire Coordination Measures. Fire control lines and bomb lines are established to protect both the link-up force and the guerrilla forces from each other's fires. Again because of the dispersion existing among guerrilla units and the fact that civilian support organizations are a part of the UW force, additional restrictions on supporting fires are necessary. In particular, the employment of nuclear and CB weapons within guerrilla warfare operational areas must be severely curtailed and thoroughly coordinated when used.

f. Communications Coordination. Generally, radio communications equipment with the guerrilla forces is severely limited. The tactical commander must provide equipment with a voice capability which can link the guerrilla force to his headquarters. This equipment is brought into the area by the liaison party. Visual recognition signals are selected to assist in link-up. In the event the necessary pyrotechnics and other markings are not available to the guerrilla force they are provided by the link-up force.

g. Employment Following Link-Up. Generally, the theater commander prescribes the conditions and duration of utilization of the guerrilla forces after link-up. Within this guidance the tactical commander may employ recovered guerrilla forces. For a discussion of employment after link-up, see paragraphs 140 through 147.

Section VI. EMPLOYMENT OF UW FORCES AFTER LINK-UP

140. General

In the event control of guerrilla forces is retained by the United States, missions may be assigned guerrilla forces after link-up with friendly forces has been accomplished. Operational control of guerrilla forces may be passed to theater army logistical command (TALOG), theater army civil affairs command (TACAComd) or retained by the tactical commander. Usually Special Forces detachments should remain with guerrilla units during this period.

141. Missions

a. Reconnaissance and security missions may be executed by guerrilla units such as screening the flanks of friendly forces; patrolling void areas between dispersed units and providing guides.

b. When properly trained, organized and supported, certain guerrilla units may have the capability of performing conventional combat operations. Normally, supporting combat units such as artillery and armor are. provided by the tactical commander. As an example, the containment or destruction of bypassed enemy units may be assigned to guerrillas.

c. Rear area security missions such as guarding supply depots, lines of communication, military installations and prisoner of war compounds may be assigned to guerrilla units.

d. Counter guerrilla operations directed against enemy dissidents may be performed by guerrilla units.

e. Guerrilla forces may be utilized to assist civil affairs units. Such tasks as police of civilian communities, collection and control of refugees and assistance, in civil administration are examples of civil affairs assistance missions to which guerrilla units may be assigned.

142. Command Relationships

a. The theater commander prescribes the conditions of employment and duration of attachment of guerrilla forces to conventional commands after link-up. Guerrilla forces may be utilized by tactical commanders or attached to other theater service components or theater army commands.

b. Guerrilla units are attached to the conventional force and responsibility for administrative support of these units passes from the SFOB to the conventional force.

c. In most situations, Special Forces detachments should remain with the guerrilla force during post link-up operations. The requirement for their employment in other operational areas, coupled with the efficiency of, and type missions assigned, are factors governing the retention of Special Forces detachments with the guerrilla force.

d. Upon completion of the mission or when directed by the theater commander, guerrilla forces are released for demobilization and return to their national government.

143. Conventional Combat Operations

Properly trained and equipped guerrilla units can be employed as conventional combat units. Normally, they require additional combat and logistical support such as armor, artillery and transportation. A period of retraining and reequipping is usually necessary prior to commitment to combat. When so employed they should be commanded by their own officers. Usually the special forces detachment remains with the guerrilla unit to assist them in the transition to the status of a combat unit operating in a strange environment under unknown higher commanders.

144. Reconnaissance and Security Missions

a. Because of their familiarity with the terrain and people in their operational areas, guerrilla forces possess a unique capability in a reconnaissance and security role. However, their lack of vehicular mobility and voice communications equipment are limitations on their employment with mobile forces. When employed with mobile units, the tactical commander may provide the necessary transportation and communications equipment for selected guerrilla units.

b. The normal method of employment in reconnaissance and security missions is to assign guerrilla units an area of responsibility (fig. 47). Within this area guerrilla forces patrol difficult terrain and gaps between tactical units, establish road blocks and observation posts, screen flanks, provide guides to conventional units and seek out enemy agents and stragglers.

145. Rear Area Security

a. Guerrilla forces may be assigned rear area security missions with various tactical commands or within the theater army logistical command area. In assigning guerrilla forces a rear area security role, their area knowledge should be the governing factor and, where possible, they should be employed within areas familiar to them.

b. They may be used as security forces at logistical and administrative installations, supply depots, airfields, pipelines, rail yards, ports and tactical unit trains areas. Guerrilla omits .,an patrol difficult terrain which contains bypassed enemy units or stragglers; police towns and cities; guard lines of communications such as railroads, highways, telecommunications systems and canals. When provided with appropriate transportation, guerrilla units may be employed as a mobile security force reserve.

c. Selected guerrilla, auxiliary, and underground elements may be effectively used in support of civil censorship operations conducted throughout the controlled area.

146. Counter Guerrilla Operations

Guerrilla forces are adapted by experience and training for use in counter guerrilla operations. Their knowledge of guerrilla techniques, the language, terrain and population are important capabilities which can be exploited by conventional commanders engaged in counter guerrilla operations. Guerrilla forces may provide the principal sources of intelligence information about dissident elements opposing friendly forces. They have the capability of moving in difficult terrain and locating guerrilla bands. They detect guerrilla supporters in villages and towns and implement control measures in unfriendly areas. When properly organized and supported, guerrilla forces may be made completely responsible for counter guerrilla operations in selected areas.

147. Civil Affairs Assistance

Because of their knowledge of the language and familiarity with the local population, guerrilla forces or selected civilian support elements may be assigned to assist civil affairs units. They may be directly attached to divisional, corps or army civil affairs units or placed under command of the theater army civil affairs

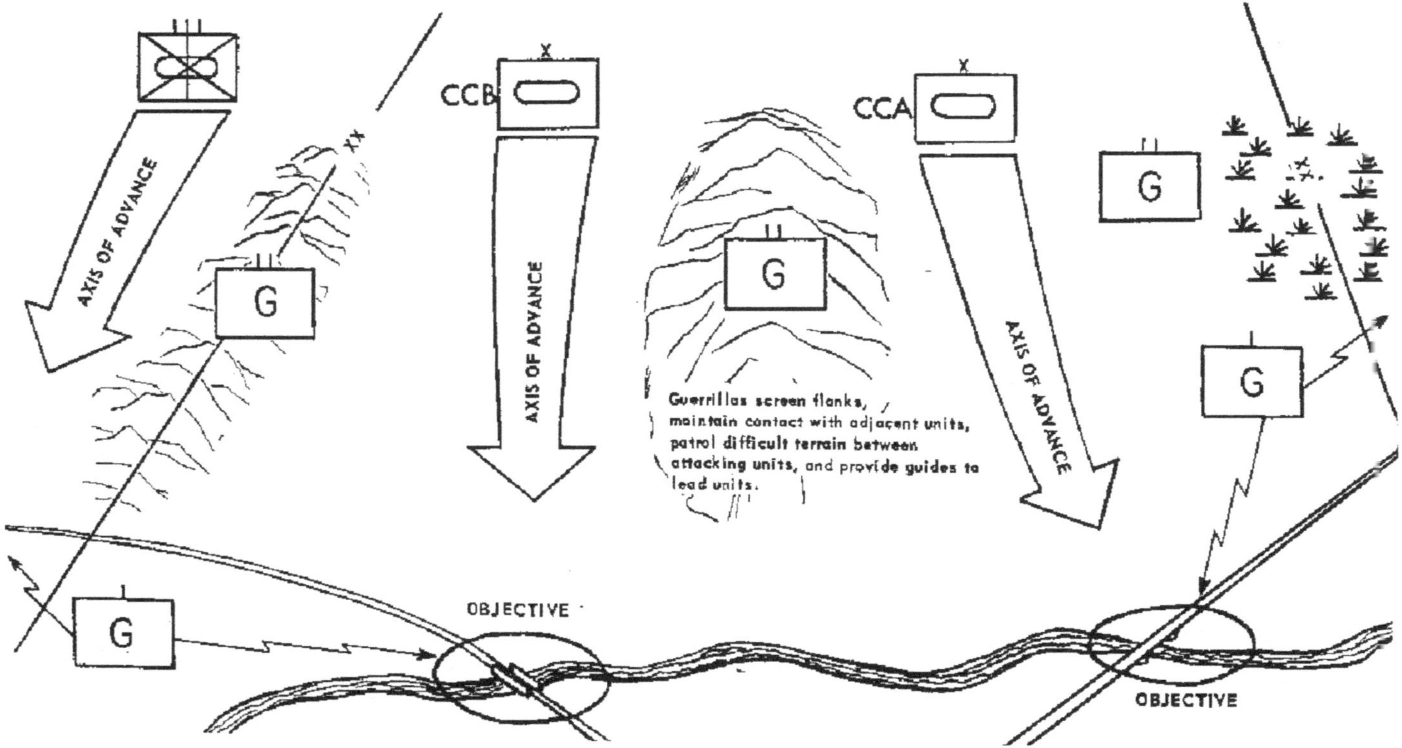

Figure 47. Employment of guerrillas in a reconnaissance and security role.

command (TACAComd). Guerrilla forces can perform refugee collection and control duties, civil police duties, assist in the psychological operations campaign in rear areas, help establish civil government, apprehend collaborators and spies, recruit labor, furnish or locate technicians to operate public utilities, guard key installations and public buildings, assist in the review and censorship of material for dissemination through public media facilities, and in general, assist in restoring the area to some semblance of normality.

CHAPTER 9

PSYCHOLOGICAL OPERATIONS IN SUPPORT OF UNCONVENTIONAL WARFARE

148. General

a. Unconventional warfare involves ideological, religious, political, and social factors which promote intense, emotional partisanship. Resistance organizations tend to attract personnel who accept violent change as a means of social action; they are motivated by hope for change. But, the fluid nature of resistance activity, the alternate, periods of isolation and combat, the surreptitious life makes resistance personnel particularly susceptible to propaganda affects.

b. The ideological and political factors associated with resistance activity create a fertile field for propaganda. Members of resistance movements are active propagandists. Hence, we find paralleling the guerrilla military effort a propaganda effort conducted by all resistance elements seeking to gain support for their movement. The relative isolation and clandestine atmosphere associated with resistance activities creates a continuing need for propaganda to support the effort.

c. In peace or war Special Forces units, by their very presence in a particular country, have a psychological impact on select military or paramilitary elements and on informed elements of the population. The image created by Special Forces personnel is molded by a multitude of factors which bear heavily on the successful outcome of the operation. These factors include tangible evidence of United States interest and support of the people by the presence of Special Forces personnel, the results of day-to-day, face-to-face meetings and an intelligent understanding of the objective and problems of the indigenous guerrilla force. The image is more favorable, however, if psychological operations techniques are used at all stages in the organization of the guerrilla units, especially in the preinfiltration stages, to prepare the potential guerrilla force and auxiliary forces for the arrival of United States personnel and, subsequently, in pointing up mutual efforts to achieve common political and military objectives. This new focus imposes additional burdens on the detachment commander, requiring him to have a detailed knowledge of psychological operations capabilities and the imagination to use them within the peculiar operational environment in which he is immersed. The psychological implications of unconventional warfare make a knowledge of psychological operations important. This is particularly true when Special Forces operations are predominantly psychological operations, such as in the initial phases of forming guerrilla units and seeking to win the assistance of supporting elements.

d. This chapter outlines how psychological operations assist Special Forces units in carrying out their missions, helping to maximize the chances for success and thereby contributing to a shortening of the conflict.

149. Concept and Organization

Planned psychological operations assist in the conduct of unconventional warfare operations both before and during hostilities and through those cold war activities in which the United States Army may be engaged. These psychological operations are designed to create, reinforce or sustain those attitudes held by the population which cause them to act in a manner beneficial to their own and to United States objectives.

a. National Programs. The United States Information Agency (USIA) conducts psychological operations which have the broad objective of generally defining American principles and aims and interpreting America, and its people to other peoples. This includes supporting the right of all of the peoples of the world to choose their own form of government. USIA programs can be used to prepare potential or designated Special Forces operational areas for the psychological acceptance of American military personnel.

b. Theater and Service Component Commands. Army psychological warfare units are available within the overseas theater or command to assist in amplifying the broad policies and goals in the particular area in which unconventional warfare units are committed. During hostilities a psychological operations staff officer coordinates with the Joint Unconventional Warfare Task Force (JUWTF) to assist Special Forces detachments in their respective areas of operations. Planning for Special Forces operations includes the use of psychological operations in all phases of the unconventional warfare operation, from psychological preparation stage through demobilization.

150. Target Audiences

a. Enemy Target Audience. The enemy target audience may consist of several. elements:

(1) Enemy Military Forces.

(a) Enemy military forces may be of the same nationality as the population or they may represent an occupying foreign power. In either case the guerrilla force and the auxiliary personnel supporting them attempt to make enemy soldiers feel isolated and under-supported by pointing up any inadequacies in their supplies and equipment, and the perennial danger of death. By focusing on the enemy soldier's frustrations, psychological operations can lower his morale and reduce his effectiveness, particularly in conjunction with the powerful pressures generated by continuous combat action. Ambushing supply columns, sniping, small scale raids against isolated units, cutting enemy communications lines and the destruction of vital objectives at night induce a basic feeling of inadequacy, insecurity and fear in the enemy soldier. This feeling of inadequacy and fear permit easy access to the mind with the several tools of psychological operations, and make the enemy soldier vulnerable to appeals urging surrender, malingering, or desertion. The enemy soldier's feeling of isolation and his receptivity to our appeals are further aided through leaflets and broadcast messages which stress the popular support of the aims of the guerrillas.

(b) The psychological "isolation" campaign may be supplemented by a more positive technique designed to elicit more readily observable reactions. If the Special Forces commander desires to induce enemy soldiers to defect or desert, satisfying and realistic goals must be introduced to attract the target audience. The enemy soldier should be told why and how he should defect and given assurances concerning his safety and welcome by the guerrilla force. When enemy soldiers are taken by the guerrillas, promises of safety and good treatment must be kept. Proof of good treatment is passed on to enemy units by photographing the soldier, having him sign leaflets, or even having him make loudspeaker appeals to his former comrades. If these techniques are unfeasible, auxiliary personnel may inform enemy units by word of mouth of the well being of defected or captured personnel. Obviously, the defection of an enemy soldier is important news to his former

colleagues, since it indicates to those remaining behind that a defector's safety is assured. This fact can have a great psychological impact on the enemy and on the guerrillas themselves—the enemy is made to feel that his own comrades, are wavering and do not support the enemy goals; while the guerrillas learn that the enemy is weakening and their own chances for success increasing.

(2) *Civilian collaborators*. Civilians in the operational area may be supporting a puppet form of government or otherwise collaborating with an enemy occupation force. Themes and appeals disseminated to this group vary accordingly, but the psychological objectives are the same as those for the enemy military. An isolation program designed to instill doubt and fear may be carried out and a positive political action program designed to elicit active support of the guerrillas also may be effected. If these programs fail, it may become necessary to take more aggressive action in the form of harsh treatment. Harsh treatment of key collaborators can weaken the collaborators' belief in. the strength and power of their military forces. This approach, fraught with propaganda dangers, should be used only after all other appeal means have failed. If used, they must be made to appear as though initiated and effected by the guerrillas to reduce the possibility of reprisals against civilians.

b. Civilian Population.

(1) No guerrilla movement can succeed without a majority of the population being favorably inclined toward it. Often, however, in the initial stage of hostilities, the population, because of fear or uncertainty about the aims of the movement, may be neutral or opposed to the guerrillas. This is understandable because the population is caught between the demands and controls of the enemy force and those of the guerrillas, In this instance, force and the main objective of psychological operations in guerrilla warfare is to persuade the target group that the guerrillas are fighting for the welfare and goals of the population, that these goals are attainable and that the United States in supporting the guerrilla force is pressing for the same political and social goals. Psychological programs aimed at this target audience stress appeals designed to induce the population to support and obey the guerrillas in achieving recognized common objectives.

(2) By their presence in the operational area, Special Forces personnel are able to gather exploitable information on the immediate situation and on the attitudes and behavior of the local population. The guerrilla force and its supporting elements are a valuable storehouse of information which can be used to strengthen psychological operations plans directed at civilian and enemy target audiences. Armed with this information, the special forces commander can then request support from the theater psychological operations officer to assist in carrying out a predetermined and coordinated psychological program. This support may take the form of dropping newspapers and other, semiofficial media to the population, supplying the guerrillas with material to produce printed Matter and providing the special forces commander with additional advice and techniques to conduct a detailed and integrated psychological program to supplement the guerrilla operation.

c. Guerrillas and the Auxiliaries. The third major target audience to be considered by the Special Forces commander comprises the guerrillas, the auxiliaries, and those underground elements assisting the guerrillas. The guerrilla force has been given proof that the United States supports the general objectives of the guerrilla movement. But, as the representative of the United States theater commander, the Special Forces detachment commander must insure that specific goals for the guerrillas and its support elements are reinterpreted and reemphasized continually during the hostilities phase.

151. Types of Psychological Warfare Operations in Guerrilla Warfare Operational Areas

a. Action Operations. Action operations are those operations taken by the Special Forces commander which are designed to have a psychological effect on any of the three major target audiences. As indicated above, some combat actions may be initiated by the Special Forces commander purely for psychological purposes, especially those related to raising the morale of the guerrilla fighters or to manifest guerrilla support of the people. The purpose of these actions is to reinforce belief in the strength of the guerrilla force and in the rightness of their goals. These beliefs when held by the population open up sources of food and information required for the survival of the guerrilla force. Enemy credence in the strength of the guerrilla force tends to lower his morale and weaken the efficiency of his operations. Examples of actions initiated primarily for psychological reasons that can be taken by the special forces commander are:

(1) Assisting the civilian population by distributing and administering medical supplies;

(2) The rescue and evacuation of key civilians supporting the guerrilla cause;

(3) Warning the civilian population of impending aircraft or missile attacks in the local area. These warnings imply guerrilla control over the operation and further increase the belief in the strength of the guerrilla force;

(4) When area supremacy is achieved, encouraging and assisting the civilian population to resume their normal activities. This may involve use of the guerrillas or auxiliary units in assisting the local population to repair buildings, build needed structures, harvest crops, reopen schools and churches, organize social activity groups, etc.;

(5) The institution of honest and effective government in the area. These psychological programs must carry the full weight of the prestige and legality of the United States and its allies. This is demonstrated by having appropriate directives emanate from United States authorities at theater level or higher. Joint directives issued by United States and indigenous guerrilla leaders or a credible government-in-exile give added force to the action programs.

(6) Meeting civilians face-to-face. During those periods of operations before the Special Forces commander can actively assist the civilian population to resume a relatively normal life, the commander must reinforce written appeals by conducting meetings or discussions with the local civilians. These provide additional tangible evidence to the population that the guerrillas are supported by the United States and that both are working in the interests of the population. Members of the Special Forces detachment participate in such meetings to establish full rapport with the population, thereby diminishing the "foreignness" of Special Forces personnel. These meetings help identify the guerillas and United States personnel with the population.

b. Printed Media. The leaflet, poster or bulletin is the most common and most effective type of printed material used by the

Special Forces commander and the guerrillas in the operational area. Small printing presses and other simple types of reproducing machines can be used to print leaflets and news communiqués. The technical problems associated with printing may be considerable and dissemination of the leaflets difficult in those areas where the enemy is able to maintain firm control. In the initial stages of hostilities, when psychological operations are most vital, guerrilla forces may not have the facilities to produce large amounts of printed material.

The techniques of leaflet writing for unconventional operations are the same as those for conventional programs. Guerrillas, aided by the Special Forces commander, can usually select themes which are more timely, more credible and more consistent than those which emanate from sources outside the operational area. The Special Forces commander can augment the locally prepared program by having' small newspapers dropped into the area to supplement bulletins issued through auxiliary unit channels. Printed material should be used to emphasize favorable aspects of civic action programs already undertaken. War aims should be publicized as aspects of permanent national aims and policies and disseminated as official looking leaflets. Leaflets carrying the official text of joint communiqués signed by the theater commander and known resistance leaders should be official and formal in appearance when issued to the target audience.

c. Rumor. Rumor can be an effective propaganda device, especially when employed to disseminate black propaganda. The Special Forces commander, using guerrillas and auxiliary information channels, can initiate rumor campaigns in the operational area, if the situation calls for them. Themes that the Special Forces commander would be reluctant to sanction as official information can be spread through the medium of rumor. Although rumors are difficult to control and the target audience never specifically isolated, this medium does have the advantage of being virtually impossible to trace. Since this device is also exploitable by the enemy, rumors which are detrimental to the guerrilla effort should be countered by leaflet or face-to-face meetings with selected members of the civilian population.

152. Psychological Operations to Support Demobilization

Psychological operations are used to assist in the demobilization of a guerrilla force. They consist of programs using all media to explain to the guerrilla steps to be taken in the demobilization process. In addition, rehabilitation programs, sponsored by the United States or the national government concerned, are explained to the guerrillas with emphasis on the guerrilla's role in the future plans for their country. In general, psychological operations aid in the orderly transition of the guerrilla force to more normal pursuits and prepare the civilian population for the return of guerrilla elements.

CHAPTER 10
DEMOBILIZATION

153. General

When juncture between friendly conventional troops and the area command is completed, the ability of guerrilla forces to support military operations gradually diminishes. Units retained beyond their period of usefulness may become a liability and a potential source of trouble. Consideration is given to the demobilization of guerrilla contingents in sectors occupied by U.S. troops. The decision regarding the transfer of guerrilla forces and associated organizations to the national government concerned is one which must be resolved at the theater level. Problems of international relationships, attitudes of the civil population toward these forces, and vice versa, and the political, economic and social implications of such a transfer are a paramount consideration. In the event that no recognized national government exists, the decision to disband the forces, in part or in their entirety, likewise requires careful consideration. Disbanding of guerrilla forces when composed of elements foreign to the area may be extremely dangerous. In any case, Special Forces units may be involved in demobilization procedures. Measures to achieve adequate coordination between special forces, civil affairs (CA) and other appropriate military and political authorities are instituted to insure a disposition of guerrilla forces in harmony with the long-range political objectives of the United States in the area.

154. Role of Sponsoring Powers

a. When a theater command has completed combat operations with a guerrilla force, it may release the force to the provisional government recognized by the United States.

b. Although the responsibility for demobilization and utilization of guerrilla forces belongs to the provisional government, the United States is responsible for restoring and maintaining public order, as far as possible, and may have to assume these obligations temporarily until an effective administration has been established.

155. Planning

a. Initiation of Plans. Long-range planning for the eventual disposition of the guerrilla force commences at theater level as soon as these forces have been organized. Planning is continuous and is revised concurrently with operations to reflect the existing political and military situation. Appropriate instructions are included in theater civil affairs plans. Decisions affecting the eventual disposition of U.S. sponsored guerrilla and associated forces are made at the highest political and military levels in the theater.

b. Civil Affairs Role. Demobilization instructions are written into CA annexes to theater plans. Also, CA teams may be provided to assist in demobilization procedures, particularly when no suitable provisional government exists to assume control. CA personnel are normally attached to Special Forces detachments prior to release of former guerrillas to CA authority in order to maintain adequate liaison throughout the transition and demobilization period.

c. Special Forces Role. Commanders of Special Forces units that have been sponsoring guerrilla units and commanders of CA elements that are assuming responsibility establish liaison to assure turnover without loss of control or influence. SF commanders provide CA commanders with the following:

(1) All available lists of guerrillas, their supporters and other key inhabitants, together with any knowledge as to their political attitudes, their leadership or administrative potential, and other information that might be helpful in operations subsequent to the UW phase.

(2) Area studies and intelligence not already available to CA elements.

156. Demobilization Courses of Action

a. Demobilization by U.S. forces may take any one or a combination of the following courses:

(1) The guerrilla forte, with all arils and equipment, may be released to the recognized government.

(2) The guerrilla force, minus U.S. supplied arms and equipment, may be released to the recognized government.

(3) The guerrilla force may be, demobilized and relocated by the U. S.

b. Demobilization is planned and conducted so as to include the following:

(1) Assembly of the guerrilla force.

(2) Completion of administrative records.

(3) Settlement of pay, allowances, and benefits.

(4) Settlement of claims.

(5) Awarding of decorations.

(6) Collection of arms and equipment.

(7) Care of sick and wounded.

(8) Discharge.

(9) Provision for the rehabilitation and employment of discharged guerrillas. Prevention of bandit or antigovernment bands forming from guerrilla elements.

157. Assembly of the Guerrilla Force

a. The guerrilla force is gathered by units into assembly areas. All records and equipment are brought with the units. Hospitals and convalescent camps are centrally located. Training programs are conducted to occupy and reorient the men.

b. The guerrilla force, during demobilization, may represent a powerful political element in the liberated area. Support from its members for various causes can be sought by factions both within and outside the guerrilla forces. In the interest of orderly demobilization, political activity by or among the, guerrillas is closely supervised and movement of the guerrillas is controlled to prevent desertions and absence without leave.

158. Completion of Administrative Records

All elements of the guerrilla force complete the administrative records of their units. Certificates are prepared to cover records that have been lost or destroyed: Complete payrolls are prepared and are reconciled with authorized unit strength figures. Arms and equipment are inventoried and accountability is established.

159. Settlement of Pay, Allowances and Benefits

Members of the force are paid after previous partial payments have been deducted. Authorized benefits are paid to legal survivors of men who have died or were killed in action.

160. Settlement of Claims

Administrative delay in the settlement of claims arising from t he activities of resistance forces is d, potential source of ill will and often results in injustice. The method of settlement outlined below eliminates the need for an elaborate claims service by a headquarters which may be required to act without adequate information. It also makes possible the prompt payment of claims and minimizes the possibility of fraud.

a. A fixed sum is credited to the recognized local government for settlement of authorized obligations incurred by guerrilla forces prior to their demobilization. Within that sum and prior to an announced future date, claims may be approved and certified to CA; the CA commander, after reviewing available records authorizes payment. The above procedure does not apply in the case of claims made against the United States.

b. Claims teams are set up within each guerrilla unit having authority to issue receipts or otherwise incur financial obligation. Disbursing officers are attached to each claims team. Notices are published in the area of operations announcing that claims teams will be present on specified dates to receive and pay claims.

c. The claims team establishes an office in the area and brings with it the records pertaining to receipts and expenditures. Receipts are verified and approved by the guerrilla members of the team and presented to the disbursing officer who makes immediate payment to the claimant from funds credited to the unit. Claims for services or damages not covered by receipts, if they are below a specified amount, are processed by the claims team based on information, available. Larger claims are forwarded to higher headquarters for action.

161. Awards and Decorations

Prompt action is taken on recommendations for decorations and awards for deserving guerrillas and other resistance members. The awards are made at local ceremonies attended, when practical, by the guerrilla troops, the civilian population, high-ranking officers of the conventional forces and officials of the provisional government as soon after an operation as possible.

162. Collection of Arms and Equipment

a. If arms and equipment are to be collected, they are turned in by the guerrillas before the settlement of pay, allowances and benefits. Care is taken that weapons are not hidden for later and unlawful use. Public announcement is made that weapons must be turned in and that, after a specified date, unlicensed Possession of weapons or military equipment will be unlawful.

b. In the event that the guerrilla force, with arms and equipment, is to be turned over to a recognized national government, this phase is omitted. Inventories of arms and equipment in hands of the guerrillas are conducted jointly by representatives of the local national government and U.S. forces.

163. Care of Sick and Wounded

Guerrilla hospitals are kept in operation until the patients can be taken over by military hospitals or by civilian institutions. Every effort is made to insure that wounded and sick guerrilla soldiers are given necessary care. Permanently disabled guerrillas may be granted pensions by the recognized government.

164. Discharge

In the event that a person sworn in as a member of a guerrilla force is to pass from control of the Special Forces detachment, that person is given a discharge and testimonial of his services. The discharge provisions applicable to military personnel will be used as a guide. Current Department of Army forms appropriately modified may be used.

165. Rehabilitation and Employment of Discharged Guerrillas

a. Suitable measures are taken to assist discharged guerrillas in assuming their places in civilian life. Some may be given employment by the conventional forces or by the newly constituted government. Individuals or entire units may be incorporated into the police or armed forces of the new government. Where feasible, assistance in rebuilding damaged houses or farms belonging to guerrillas may be granted. However, rehabilitation does not usually involve U.S. forces where a provisional government capable of rendering aid exists.

b. Perhaps the greatest danger in any demobilization program is the possibility that former guerrillas will resort to dissidence, factional quarrels or even to banditry. Others may take advantage of the prevalent unstable conditions to organize quasi-military or political groups which will conflict with the provisional government or U.S. authorities. It is vital, therefore, that demobilization procedures be executed expeditiously and with foresight. Procedures which are instituted will be an outgrowth of deliberations on a high level by military and political authorities. In the implementation of directives, maximum coordination between Special Forces, CA and other appropriate elements is necessary. To preclude troublesome situations from arising, tight control measures should be instituted and persons suspected of favoring action hostile to the established authority are kept under surveillance. Every effort is made to foster acceptance on their part of peaceful means to bring about a restoration of the governmental structure and assimilate the readjustments in society which accompany a cessation of wartime pursuits. Psychological operations can be of considerable assistance in these activities.

166. Auxiliaries and the Underground

Demobilization procedures usually have little effect upon the auxiliaries and the underground. Where possible, the area command furnishes names of known active underground and auxiliary personnel to the new government. It can be anticipated that these personnel will receive less in the way of actual benefits than the guerrillas but they should receive some public recognition for their services.

APPENDIX I
REFERENCES

1. General

JCS PUB 1	Dictionary of United States Military Terms for Joint Usage.
AR 220-50	Regiments, General Provisions
AR 320-5	Dictionary of United States Army Terms
AR 320-50	Authorized Abbreviations and Brevity Codes
FM 21-5	Military Training
FM 21-6	Techniques of Military Instruction
FM 21-30	Military Symbols
FM 21-50	Ranger Training
FM 21-75	Combat Training of the Individual Soldier and Patrolling.
FM 21-76	Survival
FM 21-77	Evasion and Escape
FM 21-77A	Evasion and Escape (U)
FM 21-150	Hand-to-Hand Combat
FM 31-21A	Guerrilla Warfare and Special Forces Operations (U) .
FM 31-30	Jungle Operations
FM 31-40	Tactical Cover and Deception (U)
FM 31-70	Basic Cold Weather Manual
FM 31-71	Northern Operations
FM 33-5	Psychological Warfare Operations
FM 100-1	Field Service Regulations Doctrinal Guidance (U)
FM 100-5	Field Service Regulations, Operations
FM 101-5	Staff Officers' Field Manual; Staff Organization and Procedure
FM 101-10	Staff Officers' Field Manual; Organization, Technical and Logistical Data, Part I.
DA Pam 108-1	Index of Army Motion Pictures, Film Strips, Slides and Phono-Recordings.

DA Pam 310-series	Military Publications Indexes (as applicable)
NWP 43	Evasion and Escape (U)

2. Demolitions and Mines

FM 5-25	Explosives and Demolitions
FM 5-31	Use and Installation of Boobytraps
FM 5-34	Engineer Field Data
FM 9-40	Explosive Ordnance Reconnaissance and Disposal
FM 20-32	Land Mine Warfare
FM 31-10	Barriers and Denial Operations
SM 9-5-1375	FSC Group 13: Ammunition and Explosives; Class 1375: Explosives, Solid Propellants, and Explosive Devices
TM 5-223	Foreign Mine Warfare Equipment
TM 9-1910	Military Explosives
TM 9-1940	Land Mines
TM 9-1946	Demolition Materials

3. Weapons

FM 23-5	U.S. Rifle, Caliber .30 M1
FM 23-7	Carbine, Caliber .30 M1, M1A1 M2, M3
FM 23-15	Browning Automatic Rifle, Caliber .30 M1918A2
FM 23-25	Bayonet
FM 23-30	Grenades and Pyrotechnics
FM 23-32	3.5-inch Rocket Launcher
FM 23-35	Pistols and Revolvers
FM 23-41	Submachine Guns, Caliber .45, M3 and M3A1
FM 23-55	Browning Machineguns, Caliber .30 M1917A1, M1919A4, M1919A4E1, M1919A6, and M37
FM 23-85	60-mm Mortar, M19
FM 23-90	81-mm Mortar and M29

4. Communications

FM 11-16	Signal Orders, Records, and Reports
FM 24-18	Field Radio Techniques
TM 11-263	Radio Set, AN/GRC-9, AN/GRC-9A, AN/GRC-9X, AN/GRC-9Y AN/GRC-9AX AN/GRC-87, and AN/VRC-34

TM 11-296	Radio Set, AN/PRC-6
TM 11-612	Radio Sets, AN/PRC-8, AN/PRC-8A, AN/PRC-9, AN/PRC-9A and AN/PRC-10 AN/PRC-10A, and AN/PRC-28
TM 11-666	Antennas and Radio Propagation
TM 11-486-6	Electrical Communication Systems Engineering, Radio.
TM 11-5122	Direct Current Generator, G•43/G
TM 32-220	Basic Cryptography (U)
ACP 121	Communication Instructions, General
ACP 122	Communication Instructions, Security
ACP 124	Communication Instructions, Radio Telegraph
ACP 131	Communication Instructions, Operating Signals

5. Medicine

FM 8-10	Medical Service, Theater of Operations
FM 8-35	Transportation of the Sick and Wounded
FM 8-50	Bandaging and Splinting
FM 21-10	Military Sanitation
FM 21-11	First Aid for Soldiers
TM 8-230	Medical Corpsman and Medical Specialist

6. Air and Amphibious Operations

a. Air Operations.

(1) *Joint Air Force/Army Publications.*

USCONARC TT 110-101-1 (TACM 55-2) Joint Airborne Operations.

USAFE Supplement to USCONARC TT 110-101-1 (TACM 55-2)

(2) *Air Force Publications.*

TACM 55-13 TAC Standardization Manual, Troop Carrier Aircraft.

UTS 120-4 Troop Carrier Units (Medium) (Assault)

(3) *Air National Guard Publications.*

CONAC Aircrew Training Handbook 200-4, Air National Guard, Troop Carrier Units, Medium, SA-16, Part 2.

(4) *Naval Publications.*

LFM-24 Helicopter Operations (U)

NWIP 41-6 Helicopter Operations

(5) *Army Publications.*

C5, TM 10-500 Air Delivery of Supplies and Equipment: General

DA Logistics Directive No. 163-700 (U) , dated 1 June 1959.

ST 57-150 Army Pathfinder Operations, USAIS

b. Amphibious Operations.

(1) *Joint Landing Force Manuals.*

JLFM-15 (FM 110-115) Amphibious Reconnaissance

(2) *Marine Corps Landing Force Manuals.*

LFM-1	Training
LFM-2	Terrain, Hydrography and Weather
LFM-4	Ship to Shore Movement
LFM-19	Special Landing Operations (U)

(3) *Naval Warfare and Warfare Information Publications.*

NWIP 1-1	Missions and Capabilities of US Navy Ships and Aircraft (U) .
NWP 22	Amphibious Operations
NWIP 22-4	Underwater Demolition Teams in Amphibious Operations
NWIP 22-6	Ship to Shore Movement (U)
NWP 23	Submarine Operations (U)
NWP 37	Search and Rescue

7. Intelligence and Security

AR 380-5	Safeguarding Defense Information
AR 380-8	Security Classification—Special Forces Activities.
AR 381-25	Army Intelligence Collection Instructions
AR 381-205	Procedures Facilitating Intelligence Exploitation of Captured Enemy Personnel
FM 19-40	Handling Prisoners of War
FM 30-5	Combat Intelligence
FM 30-7	Combat Intelligence-Battle Group, Combat Command and Smaller Units
FM 30-9	Military Intelligence Battalion, Field Army
FM 30-15	Intelligence Interrogation (U)
FM 30-16	Technical Intelligence (U)
FM 30-19	Order of Battle Intelligence
FM 30-28	Armed Forces Censorship (Army)

FM 110-101 Intelligence Joint Landing Force Manual

DA Pam 30-102 Intelligence Collection Guide, Identification of SMD

DA Pam 21-81 Individual Training in Collecting and Reporting Military Information

DA Pam 30-26 A Guide to the Collection of Technical Intelligence

DA Pam 30-100 Intelligence Collection Guide, Telecommunications

APPENDIX II

CATALOGUE SUPPLY SYSTEM

1. General

a. This appendix is a guide for Special Forces commanders and staffs in the planning and preparation of a catalogue supply system. In addition to the catalogue, it provides information relative to packaging, rigging, and requesting procedures.

b. The catalogue supply system

(1) Utilizes a brevity code in which a single item or several associated items are identified by a code word.

(2) Comprises both packages of associated individual items and units comprising several packages. This combination permits the user maximum flexibility in choice of supplies consistent with transmission security.

(3) Is based upon the guerrilla organization described in current DA doctrine.

c. The catalogue supply system shown in this appendix is a sample only. Special forces group commanders should not hesitate to modify the basic list to conform to varying operational conditions, equipment changes, and differences in signal cryptographic systems. The catalogue provided to the operational detachment should be simplified and reproduced in miniature. Laminated cards or 35-min film rolls are suggested.

2. Packaging and Rigging

a. The packaging system is based on man-portable packages weighing approximately 50 pounds. This facilitates the removal of supplies from a reception site by carrying parties if . other transportation is not available to handle delivery containers intact. For a manageable load, the man-portable package is equipped with carrying straps or mounted on a packboard. Each package is waterproof to permit open storage.

b. The man-portable package is suitable for use in the present aerial delivery containers as well as those now under development. This type of package permits the present containers to be adapted for delivery by any means the supporting agencies may make available.

c. Weights used are approximate and are computed without packaging material and with the items stripped of shipping containers to their inside weatherproof covering, where applicable.

d. Separate clothing packages for Special Forces personnel are omitted. In general, Special Forces personnel draw clothing from supplies issued for guerrilla use. In the event that separate special clothing packages are required for Special Forces personnel, these may be added to the catalogue in the theater of operations.

e. The efficiency of the catalogue supply system is increased by use of the following procedures

(1) Packing, with all equipment, instructional material which is printed in the appropriate language. Such material is simply written, confined to essentials, and makes the maximum use of graphics.

(2) Inclusion of an inventory list in each delivery container to aid in identification of lost or damaged material.

(3) Maximum use of reusable items for packaging material. Examples are clothing and blankets as padding and ponchos as waterproofing.

(4) Inclusion of morale and barter items which may be used to promote good will with the indigenous population or for the procurement of supplies and services.

(5) Marking each individual bundle with luminous tape or paint so that the contents are readily identified without opening the package.

f. The preparation of equipment for the various sized units is the responsibility of the Special Forces operational base. The number of delivery containers is determined by the delivery means available. The 50-pound package is utilized to the maximum in the preparation of the individual delivery containers.

3. Request Procedure

a. The code used in the catalogue supply system is a type code only and should be changed and classified when used operationally. The coding system is not secure by itself, but will reduce message length when a variety of supplies are ordered. For this example, each general type of supply is assigned letter designations:

Section	Code Designators
I —Chemical	ALHPA ALPHA through DELTA ZULU
II —Demolitions/Mines	ECHO ALPHA through HOTEL ZULU
III —Medical	INDIA ALPHA through LIMA ZULU
IV —Weapons/Ammunition	MIKE ALPHA through PAPA ZULU

Section	Code Designators
V —Quartermaster	QUEBEC ALPHA through TANGO ZULU
VI —Signal	UNIFORM ALPHA through WHISKEY ZULU
VII—Special	X-RAY ALPHA through ZULU ZULU

b. To reduce unreadable garbles when ordering supplies, use phonetic spelling. Some units and packages are followed by a numbered list showing the contents of the package or unit. For these items, the unit or package can be ordered complete, or any numbered item may be ordered separately. For example clothing and equipment for 40 men is required. Determine the boot sizes needed and include in the message. Assume that the following boot sizes are desired: Ten pair size 81/2W, six pair size 9M, three pair size 91/2N, four pair size 10N, six pair size 10M, two pair size 10W, five pair size 10 1/2M, four pair size 11M. The message would read:

ONE QUEBEC ALPHA PD BOOTS TEN SIZE EIGHT PT FIVE WHISKEY SIX SIZE NINE MIKE THREE SIZE NINE PT FIVE NOVEMBER FOUR SIZE TEN NOVEMBER SIX SIZE TEN MIKE TWO SIZE TEN WHISKEY FIVE SIZE TEN PT FIVE MIKE FOUR SIZE ELEVEN MIKE.

Clothing is packed to approximately match boot sizes (section V). On the other hand if only 40 ponchos were desired, the request would read—TWO ZERO QUEBEC ALPHA SEVEN.

c. Items listed in each unit may be ordered separately if necessary. When practical order the complete unit.

d. For items not listed, order by name in sufficient detail to identify the item. For example—TWO GASOLINE LANTERNS.

Section I. CHEMICAL

Code	Unit Designation	Unit wt	Unit data	
			No. pkgs	Contents
ALPHA ALPHA	Chemical Grenade No. 1 (16 rds).	46 lbs	1	Sixteen grenades, hand, smoke WP, M15 packed in individual con-tainers.
ALPHA BRAVO	Chemical Grenade No. 2 (16 rds).	47 lbs	1	Sixteen grenades, hand, incendiary, (TH) AN, M14 packed in indi-vidual containers.

Code	Unit Designation	Unit wt	Unit data	
			No. pkgs	Contents
ALPHA CHARLIE	Chemical Grenade No. 3 (16 rds).	34 lbs	1	Sixteen grenades, smoke, colored, M18 (Green, red, violet and yellow) packed in individual containers.
ALPHA DELTA	Chemical Grenade No. 4 (16 rds).	35 lbs	1	Sixteen grenades, hand, tear, CS, M7A1 packed in individual containers.
ALPHA ECHO	Detector kits (8).	43 lbs	1	Eight detector kits, chemical agent, M18.
ALPHA FOXTROT	Food Testing Kits (24).	45 lbs	1	Twenty-four food testing and screening kits, chemical agents, ABC-M3.
ALPHA GOLF	Leather Dressing (96).	43 lbs	1	Ninety-six cans leather dressing, vesicant gas resistant, M2.
ALPHA HOTEL	Protection & Treatment Set (70).	50 lbs	1	Seventy protection and treatment sets, chemical warfare agents, M5AL
ALPHA INDIA	Water testing Kits (24).	50 lbs	1	Twenty-four water testing kits chemical agents, AN-M2.
ALPHA JULIET	DANC Unit (1).	59 lbs	1	DANC solution unit, 3 gallon M4.
ALPHA KILO	Decontaminating Agent (1).	61 lbs	1	Decontaminating Agent, STB.
ALPHA MIKE	Protective Mask (10).	44 lbs	1	Ten masks, protective, field, M17.
ALPHA NOVEMBER	Impregnating Set (1).	57 lbs	1	1. Impregnating set, clothing, field M3 (55 lbs). 2. Kit. testing, impregnite in clothing, Ml (2 lbs).
ALPHA OSCAR	Napalm	42 lbs	1	Eight cans chemical agent thickener 5¼ lb can.

Section II. DEMOLITIONS AND MINES

Code	Unit Designation	Unit wt	Unit data	
			No pkgs	Contents
ECHO ALPHA	Demolitions No. 1 (20 Blocks).	50 lbs	1	20 blocks, demolition, M5A1 (2 ½ lb comp C-4)
ECHO BRAVO	Demolitions No. 2 (2 assemblies).	44 lbs	1	Two assemblies, demolition M37 (2 ½ lb comp C-4) 8 blocks per assembly.
ECHO CHARLIE	Demolitions No. 3 (45 blocks).	45 lbs	1	45 blocks, demolition, (1 lb TNT).
ECHO DELTA	Detonating Cord (6000 ft). Note 1	42 lbs	1	6000 ft cord, detonating, 1000 ft per spool (6 spools-42 lbs).
ECHO ECHO	Detonators (150).	45 lbs	1	150 detonators, friction, 8 second delay M2 and 15 second delay Ml packed 10 per box (15 boxes-45 lbs).
ECHO FOXTROT	Firing Device No. 1 (200).	40 lbs	1	200 firing devices, set, demolition, delay type, Ml packed 10 per box, consisting of- 1. Two 15-minute delay. 2. Three 1-hour delay. 3. Three 2 ½-hour delay. 4. One 11 ½-hour delay. 5. One 13 ½-hour delay. (20 boxes-40 lbs).
ECHO HOTEL	Firing Device No. 2 (116).	44 lbs	1	116 firing devices, demolition, mixed, packed 29 per box consisting of— 1. Five pressure type M1A1 2. Five release type M5. 3. Five pull friction type M2. 4. Five pull release type M3. 5. Five pull type M1.

Code	Unit Designation	Unit wt	Unit data	
			No pkgs	Contents
				6. Four detonators, concussion type E Ml. (4 boxes-11 lbs per box).
ECHO INDIA	Fuze (27,000 ft) *Note 1*	45 lbs	1	27,000 ft fuze, blasting, time, 100 ft, packages packed 30 packages per metal can. (9 cans-45 lbs).
ECHO JULIET	Fuze Igniters	45 lbs	1	225 igniters, blasting, fuze weatherproof M2 packed 5 per box (45 boxes-45 lbs).
FOXTROT ALPHA	Priming Material No. 1.	47 lbs	1	1. 250 caps, blasting, special type II J2 PETN packed 50 per box (5 boxes-5 lbs). 2. 6000 ft cord, detonating 1000 ft per spool (6 spools-42 lbs).
FOXTROT BRAVO	Priming Material No. 2.	48 lbs	1	1. 500 caps, blasting, special, non-electric type I J1 PETN packed 50 per can (10 cans-5 lbs). 2. 4000 ft cord, detonating, 1000.ft per spool (4 spools-28 lbs). 3. 9000 ft fuze, blasting, time, 100 ft packages packed 30 packages per metal can (3 cans-15 lbs).
FOXTROT CHARLIE	Non-electric Demolition Unit No. 1. *Note* 1	55 lbs	1	1. One assembly, demolition, M37 (22 lbs). 2. 10 blocks, demolition, 1 lb TNT (10 lbs). 3. 50 caps, non-electric (½ lb). 4. 1000 ft cord, detonating (7 lbs). 5. Two crimpers, cap. 6. 25 destructors, explo-

Code	Unit Designation	Unit wt	Unit data	
			No pkgs	Contents
				packed 5 per box (5 boxes-6 ¼ lbs). 7. 3000 ft fuze, blasting, time (1 can-5 lbs). 8. 15 igniters, blasting, fuze M2 packed 5 per box (3 boxes-3 lbs). 9. 1 roll insulation tape, electrical (1 lb).
FOXTROT DELTA	Non-electric Demolition Unit No. 2.	745 lbs	15	1. One non-electric demolition unit No. 1. 2. 250 caps, blasting, special, non – electric Type I J1 PETN packed 50 per can. 3. One detonating cord unit. 4. One time fuze unit. 5. Twelve demolition units No. 1.
FOXTROT ECHO	Electric Demolitions Unit No. 1.	93 lbs	2	1. 50 caps, electric (1lb). 2. 1000 ft cord, detonating (7 lbs). 3. 500 ft cable, power electrical firing on reel RL39B (36 lbs). 4. One assembly, demolition M37 (22 lbs). 5. 10 blocks, demolition 1 lb TNT (10 lbs). 6. 25 destructors, explosive universal M10 packed 5 per box (5 boxes-6 ¼ lbs). 7. One galvanometer, blasting (21bs). 8. One machine, blasting, 10 cap capacity (5 ½ lbs). 9. One pair pliers, lineman's (1 ¼ lbs). 10. 200 ft wire, electrical annunicator or reel (2 ¼ lbs).

Code	Unit Designation	Unit wt	Unit data	
			No pkgs	Contents
FOXTROT FOXTROT	Electric Demolition Unit No. 2.	740 lbs	14	1. One electric demolition unit No. 1. 2. 250 caps, blasting, special, electric Type II J2 PETN packed 50 per box. 3. One detonating cord unit. 4. Twelve demolitions units No. 1.
GOLF ALPHA	Anti-tank Mines No. 1. *Note 2*	50 lbs	1	10 mines, light ATM7A2.
GOLF BRAVO	Anti-tank Mines No. 2. *Note 2*	56 lbs	1	2 mines, AT, M19.
GOLF CHARLIE	Anti-Personnel Mines, No. 1. *Note 2*	40 lbs	1	128 mines, AP, M14.
GOLF DELTA	Anti-Personnel Mines, No. 2. *Note 2*	47 lbs	1	6 mines, AP, M16, bounding.
GOLF ECHO	Anti-Personnel Mines, No. 3. *Note 2*	45 lbs	1	15 weapons, AP, M18, Claymore.

NOTES:

1. Assembled in two packages due to bulk.
2. Fuzes included in each package.
3 General.
   a. Accessory items such as priming adapters and detonating cord clips may be added to the package as desired.
   b. Peculiar non-standard items are added to the list in the theater of operations.

Section III. MEDICAL

Code	Unit designation	Unit wt	Unit data	
			No. Pkgs	Contents
INDIA ALPHA	Combat Aidman's Set.	48 lbs	1	Eight surgical instrument and supply sets, individual (6 lbs) standard medical supply set

Code	Unit designation	Unit wt	Unit data	
			No. Pkgs	Contents
				consisting of— 1. One bottle of Acetyl-salicylic Acid tablets, USP bottle, 100 per bottle. 2. Five morphine in-jections, USP 16mg (¼ gr.) pkg. 3. One tube of Tectra-caine Ophthalmic ointment 1/8 oz tube. 4. One package Benzalkonium Chloride Tincture. 5. Two bandages, gauze, 3 inch. 6. Two bandages, muslin. 7. Two dressings, first aid, field, 7 ½" x 8". 8. Eight dressings, first aid, field, 4" x 7". 9. One spool of adhesive plaster, surgical, 3". 10. Two packages of bandages, absorbent, adhesive, 18 per package. 11. One pair scissors, bandage, angular, heavy. 12. One tourniquet. 13. One thermometer. 14. One card of pins, safety, 12 per card. 15. One pencil. 16. One surgical instru-ment set, minor surgery: a. Two needle holders. b. Two forceps. c. One blade handle. d. Two packages of blades No. 10, 6 per package. e. Two packages of blades No. 11, 6 per package. f. One probe. g. One pair scissors.

Code	Unit designation	Unit wt	Unit data	
			No. Pkgs	Contents
				with needles.
INDIA BRAVO	Field Surgery Set *Note 1*	50 lbs	1	1. Two bags, canvas, M-5 with shoulder straps. 2. Two holder, suture needle, 7". 3. Four forceps, hemostat, curved, 6 ¼". 4. Four forceps, hemostat, straight, 6 ¼". 5. Two forceps, hemostat, straight, 7 ¼". 6. Two forceps, Foerster type. 7. Two forceps, tissue, tweezers, 5 ½". 8. Two handle, surgical knife, No. 3. 9. Two packages, surgical knife blade No. 10. 10. Two packages, surgical knife blade, No. 11. 11. Two scissors, band-age, angular, heavy, 12. Two scissors, band-age, angular Lister, 8 13. Two scissors, general surgical curved, 7 ¼". 14. Two scissors, general surgical, straight, 7". *Sterilization and Surgical Preparation* 15. Eight containers of detergent, surgical, 5 oz. 16. Two razors, safety. 17. Six packages of blades, safety razor, 5 per package. 18. Six cakes of surgical soap 4 oz. cake. 19. Two brushes, scrub.

Code	Unit designation	Unit wt	Unit data	
			No. Pkgs	Contents
				20. Two pairs of gloves, surgical, size 7 ½". 21. Two pairs of gloves, surgical, size 8. 22. One sterilizer, surgical instrument, boiling type. 28. One heater, surgical dressing, sterilizer. 24. Four hand towels. *Suture Material* 25. Two packages of needles, suture, cat-gut, Mayo. 26. Two packages of needles, suture, surgeon's regular, 3/8 circle, cutting edge, size 2, 6 per package. 27. Two packages of needles, suture, reg-ular, %s circle, cut-ting edge, size 16, 6 per package. 28. Two spools of suture, non-absorbable, surgical silk braided, size 000. 29. Two spools of suture, non-absorbable, surgical silk braided, size 00. 30. Two spools of suture, non-absorbable, surgical silk braided, size 0. 31. Two packages of suture, absorbable, surgical, boilable, plain size 000, single armed, ½ circle needle, 12 per pack-age. 32. Two packages of suture, absorbable, surgical, boilable,

Code	Unit designation	Unit wt	Unit data	
			No. Pkgs	Contents
				armed, ½ circle needle; 12 per pack-age. 33. Two packages of suture, nonabsorb-able, surgical silk, single armed, 3/8 cir-cle needle size 000, 12 per package. 34. Two spools of tanta-lum wire, size .028. *Syringes and Needles* 35. Two syringes, 2 cc, Luer. 36. Two syringes, 10 cc, Luer. 37. Two syringes, 30 cc, Luer. 38. One box of needles, gau ge 23, 12 per package, 39. One box of needles, gau ge 20, 12 per package. 40. One box of needles, gau ge 17, 12 per package. *Anesthesia* 41, Four bottles of Lido-caine, 2%, 20 cc bottle. 42. Two bottles of Pro-caine Hydrochloride, 1%, 30 cc bottle. 43. Two ether masks. 44. Four cans of ether, 1/4 lb per can. 45. Four vials, 0.5 gm., Thiopental sodium. 46. Two tubes, petrola-tu m, 1/z oz. 47. Two bottles, atro-pine, 25 cc. 48. Two airway, phar-yngeal, Gu edal, rub-ber, adu lt. 49. One airway, phar-yngeal, Gu edal, rub-ber, child.

Code	Unit designation	Unit wt	Unit data	
			No. Pkgs	Contents
				50. Catheter, intratracheal, Magill, with rubber. *Diagnostic Instruments* 51. Four thermometers, clinical oral. 52. One stethoscope, combination type. 53. One otoscope, opthlmoscope set with batteries. 54. One blood pressure cuff, aneroid. *Bandages And Dressings* 55. Twenty-four band-ages, gauze, 3" x 6 yds, sterile. 56. Twelve bandages, muslin. 57. Two rolls, cotton purified, 1 oz. 58. Two surgical dress-ing, field, large. 59. Six surgical dress-ing, field medium. 60. Four packages, petrolatum gauze, 3 per package. 61. One package, gauze pad, 2 x 2", 199 per package. 62. One package, gauze pad, 4 x 4", 200 per package. 63. Two boxes, bandage, absorbent, adhesive, 100 per package. *Drugs and Reagents* 64. Two bottles, epinephrine, injection 30 cc. 65. Six packages, morphine syrettes, 16 mg., 5 per package. 66. Two bottles, benzalkonium chloride, 4 oz.

Code	Unit designation	Unit wt	Unit data	
			No Pkgs	Contents
				thonium chloride, 250 mg, 80 tablets per bottle. 68. One can, benzoin tincture, 1 pint, 69. Four tubes bacitracin opthalmic ointment, 1/s oz tube. 70. Four tubes, tetra-caine opthalmic oint-ment, 1/8 oz tube. 71. Eight syrette, atro-pine injection, 2 mg. *Miscellaneous* 72. Two tourniquet, non-pneumatic, 1½" x 42". 73. Two catheter, French, urethral, size 16. 74. Two catheter, Foley, urethral, indwelling size 16. 75. Two tubes, Levin, K-10 (Baxter). 76. Two adapters, Luer syringe for rubber tubing. 77. Two drains, Penrose, 3/8" x 36". 78. Four pencils, gra-phite, No. 2. 79. One book, record. 80. Two books, emerg-ency medical tag. 81. Four cards, pin, safety, orthopedic, medium, 12 per card. 82. Fifty envelopes, drug dispensing. 83. One tube, lubricant, surgical, 4 oz. 84. Two cannulas, tra-cheotomy, nylon, size 5
INDIA CHARLIE	Field Surgery Set Supple-ment.	45 lbs	1	Standard medical supply set 6545-927-4400. 1. One bag, water proof, signal equipment with

Code	Unit designation	Unit wt	Unit data	
			No. Pkgs	Contents
				x 18". 2. Two bottles of Dex-tran injection, 500 cc. 3. Six packages of Morphine injection USP 16 mgm (¼ gr), 5 per pkg. 4. One bottle of Peni-cillin C tablets USP, 250,000 units, 100 per bottle. 5. Five packages of Penicillin injection in oil 400,000 units cartridge-needle unit 1 cc, 10 per package. 6. Two syringes, car-tridge. 7. Six bottles of Al-bumin, normal hu-man serum, 100 cc. *Note 1.* 8. One box of Nalor-phine hydrochloride injection USP, 5 mgm, (1/12 gr) per cc, 6 per box. 9. Six bandages, gauze, camouflaged, 3 in x 6 yds. 10. Four bandages, muslin. 11. Two dressings, first aid, field, 11 ¾" square-large. 12. 24 dressings, first aid, field medium. 13. 36 dressings, first aid, field small. 14. One spool of ad-hesive plaster, surgi-cal, 3".
INDIA DELTA	Medical Dis-pensary Set.	40 lbs (—)	1	1. One bottle of Chlor-pheniramine maleate (chlortrimenton) 4 mg, tablet. 1000 Per bottle. 2. One bottle of Terpin hydrate 1 lb bottle. 3. One box of Tetra-hydrozoline hydro-chloride (Tyzine) nose drops, 12 per box. 4. One bottle of APC,

Code	Unit designation	Unit wt	Unit data	
			No. Pkgs	Contents
				bottle. 5. Six bottles of Co-deine, ½ gr. tablet, 100 per bottle. 6. Six packages of Morphine, syrette, 15 mg, 5 per package. 7. One bottle of Acetyl-salicylic acid (aspi-rin), 5 grain tablet, 1,000 per bottle, 8. Six bottles of Deca-vitamins, 100 per bottle. 9. Six bottles of Chlo-roquin, 500 mg tab-let, 100 per bottle. 10. Six bottles of Phenobarbital, 30 mg tablet, 100 per bottle. 11. Six bottles of Tetra-cycline, 250 mg tab-let, 100 per bottle. 12. Six packages of penicillin, procaine, 400,000 unit car-tridge, 10 per pack-age. 13. Six cans of foot powder. 14. 12 tubes of Unde-cylenic acid ointment, 2 oz tube. 15. Two boxes of Baci-tracinneomycin oint-ment, ½ oz tube, 6 per box. 16. Sift tubes of Zinc oxide, 1 oz tube. 17. One bottle of Cala-mine USP, 1 lb bottle (in powder form). 18. Two bottles of Po-tassium permanga-nate tablets, 5 grains, 100 per bottle. 19. One can of Methyl salicylate (oil of wintergreen), 1 lb can. 20. Six bottles of Bis-

Code	Unit designation	Unit wt	Unit data	
			No. Pkgs	Contents
				5 grain tablet, 100 per bottle. 21. One bottle of Opium, tincture, USP ¼ pint. 22. One bottle of Cascara sagrada, tablets, 100 per bottle. 23. Two boxes of alumi-num hydroxide magnesium tricilicate tablets, 100 per box. 24. Two bottles of propantheline bromide, 15 mg, 100 per bottle. 25. Sixteen syrettes, Atropine injection, 2 mg. *Miscellaneous* 26. Four plastic bottles of Phisohex, 5 oz. 27. One cartridge syringe. 28. One package of gauze pad, 4 x 4. 29. One stethescope, combination type. 30. Two boxes of absorbent-adhesive bandage (bandaid), 100 per box. 31. One box of tongue depressors, 100 per box. 32. Four pencils, graphite, No. 2. 33. Two packages of en- velopes, drug dis- pensing, 100 per package. 34. One book, record, ruled.
INDIA ECHO	Dental Unit	21 lbs	1	Three dental kits, emer-gency field (645-927-8440).

Code	Unit designation	Unit wt	Unit data	
			No Pkgs	Contents
INDIA FOXTROT	Splint Set	26 lbs	1	1. Splint Set consists of— a. One roll, splint set, telescopic splints, empty. b. 18 bandages, muslin. c. Three straps, leg traction. d. Three splints, wood. *e.* Three splints, leg, Thomas, half ring, aluminum. f. Three litter bars, leg splint supporting. g. Three splint supports and footrest, leg. h. Three splints, wire, ladder. i. Three rods, irrigator supporting. 2. Two blankets. 3. Six dressings, field, medium. 4. Six dressings, field, large.
INDIA GOLF	Water Pufification Unit *Note 2*	23 lbs	1	300 bottles of water purification tablets, iodine, 50 per bottle.
INDIA HOTEL	Insect Control Unit *Note 2*	25 lbs	1	Insect repellant, dretnyetolu amide.
INDIA INDIA	Surgical Prep Set	45 lbs	1	1. Twelve cakes of soap, surgical, 4 oz cake. 2. 24 containers of detergent, surgical, 5 oz. container. 3. Six hand brushes, surgical. 4. Two bowls, aluminum.

Code	Unit designation	Unit wt	Unit data	
			No. Pkgs	Contents
				zalkoniu m chloride, 25 cc bottle. 6. Twelve pair of gloves, surgical size 7 ½. 7. Twelve pair of gloves, su rgical, size 8. 8. One box of talc USP, lb box. 9. Twelve packages of cotton, sterile, 1 oz package. 10. Two safety razors. 11. Two packages of razor blades, dou ble edge. 12. Twelve hand towels. 13. Two bath towels. 14. Six lap sheets, small 15. Six surgical drapes, small.
INDIA JULIET	Narcotics Unit	20 lbs	1	1. 120 boxes Morphine sulfate, 15 mg syr-ette, 5 per box. 2. Six boxes of Nalor-phine hydrochloride injection, six per box. 3. 24 bottles of Codeine phosphate, 30 mg tablet, 100 per bottle. 4. Six bottles of Mepe-ridine hydrochloride, small.
INDIA KILO	Local Anesthesia	25 lbs	1	1. Ten cans of Pro-caine, cartridge, 25 per can. 2. Two cartridge syr-inges. 3. 20 bottles .of Lido-caine,. 2% , 20 cc bottle. 4. Sic .Syringes, 10 cc.

Code	Unit designation	Unit wt	Unit data	
			No. Pkgs	Contents
				cc. 6. Three boxes of needles, gauge # 23, 12 per box. 7. Three boxes of needles, gauge #22, 12 per box. 8. One box of needles, gauge #20, 12 per box. 9. Two cans of alcohol, USP, 1 pint can. 10. Six plastic containers of phisohex, 5 oz container. 11. One package of cot-ton, rolled.
INDIA LIMA	Minor Burns Unit	40 lbs	1	1. Twelve detergent, plastic container, surgical. 2. Two aluminum bowls. 3. 30 packages of pet-latum gauze packs, 12 *per* package. 4. Two packages of gauze pads, 2 x 2. 5. Two packages of gauze pads,. 4 x 4. 6. Two boxes of band-ages, gauze, sterile, 3 in, twelve per box. 7. Two boxes of band-ages, gauze, sterile, 4 in., twelve per box. 8. Two boxes of elastic bandages, 3 in., 12 per box. 9. One box of elastic bandages, 4 in:, 9 per box. 16. One jar. of Nitrofu-azone, 1 lb jar. 11. Six hand towels.
INDIA MIKE	Major Burns Unit	45 lbs	1	1. Six units of albumin, 100 cc unit.

Code	Unit designation	Unit wt	Unit data	
			No. Pkgs	Contents
				Note 1. 2. Three units of Dextran, 500 cc unit. 3. Four bottles of normal saline, 1000 cc bottle. 4. Four bottles of Dextrose, 5%, 1000 cc bottle! 5. Ten intravenous tub-ing assemblies. 6. 24 packages of sodium chloridebicarbonate mix, 2 per package. 7. Twelve boxes of procaine penicillin 400 , 000 units cartridge field type, 10 per box. 8. Two cartridge syringes. 9. Two Catheters, my dwelling, French, size 16. 10 , Two Catheters, indwelling, French, size 18. 11. One Fluid calcu lator for bu rns (non-standard). 12. Six packages of Morphine injection 15 mg, 6 per package.
INDIA KILO	Local Anesthesia	25 lbs	1	1. Ten cans of Procaine, cartridge, 25 per can. 2. Two cartridge syringes. 3. 20 bottles of Lidocame, 2%, 20 cc bottle. 4. Six syringes, 10 cc. 5. Three syringes, 20 cc. 6. Three boxes of needles, gauge # 23, 12 per box.

Code	Unit designation	Unit wt	Unit data	
			No. Pkgs	Contents
				7. Three boxes of needles, gau ge # 22, 12 per box. 8. One box of needles, gau ge #20, 12 per box. 9. Two cans of alcohol, USP, 1 pint can. 10. Six plastic containers of phisohex, 5 oz container. 11. One package of cot- ton, rolled.
INDIA LIMA	Minor Burns Unit	40 lbs	1	1. Twelve detergent, plastic container, surgical. 2. Two aluminum bowls. 3. 30 packages of petrolatum gauze packs, 12 per package. 4. Two packages of gauze pads, 2 x 2. 5. Two packages of gauze pads, 4 x 4. 6. Two boxes of bandages, gauze, sterile, 3 in, twelve per box. 7. Two boxes of bandages, gauze, sterile, 4 in., twelve per box. 8. Two boxes of elastic bandages, 3 in., 12 per box. 9. One box of elastic bandages, 4 in., 9 per box. 10. One jar of Nitrofurazone, 1 lb jar. 11. Six hand towels.
INDIA MIKE	Major Burns Unit	45 lbs	1	1. Six units of albumin, 100 cc unit, *Note 1.* 2. Three units of Dextran, 500 cc unit. 3. Four bottles of nor-

Code	Unit designation	Unit wt	Unit data	
			No. Pkgs	Contents
				bottle. 4. Four bottles of Dextrose, 5%, 1000 cc bottle. 5. Ten intravenous tub-ing assemblies. 6. 24 packages of sodium chloridebicarbonate mix, 2 per package. 7. Twelve boxes of procaine penicillin 400,-000 units cartridge field type, 10 per box. 8. Two cartridge syringes. 9. Two Catheters, indwelling, French, size 16. 10. Two Catheters, indwelling, French, size 18. 11. One Fluid calculator for burns (non-standard). 12. Six packages of Morphine injection 15 mg, 6 per package. 13. Six cannulas, tracheotomy, nylon, size 5.
INDIA NOVEMBER	Shock Set No. 1. Note 3	30 lbs	1	24 bottles of albumin, normal human serum, 100 cc bottle with injection assembly.
INDIA OSCAR	Shock Set No. 2.	24 lbs	1	12 bottles of Dextran, 500 cc bottle with injection assembly.
INDIA PAPA	Shock Set No. 3.	25 lbs	1	1. Three bottles of normal saline, 1,000 cc bottle. 2. Three bottles of Dextrose in water, 5%, 1,000 cc bottle.

Code	Unit designation	Unit wt	Unit data	
			No. Pkgs	Contents
				injection assemblies. 4. One spool of adhesive tape, 3 in. 5. Two boxes of Levarterenol ampule, 12 per box. 6. Two vials Epinephrine 1:1000 30 cc vial.
INDIA. QUEBEC	Intravenous Fluids No. 1.	25 lbs	1	1. Four bottles of Dextrose in water 5%, 1000 cc bottle. 2. Two bottles of Dextrose in water 10%, 100 cc bottle. 3. One spool of adhesive tape, 3 in. 4. Eight intravenous tubing assemblies.
INDIA ROMEO	Intravenous Fluids No. 2.	25 lbs	1	1. Six bottles of normal saline, 1,000 cc bottle. 2. Eight intravenous tubing assemblies. 3. One spool of adhesive tape, 3 in. 4. One box of Potassium chloride 10 cc vial, six per box.
INDIA SIERRA	Intravenous Fluids No. 3.	25 lb	1	1. Six bottles of sodium lactate, 1,000 cc bottle. 2. Eight intravenous tubing assemblies.
INDIA TANGO	Sterile Dressing Unit	26 lbs	1	1. Twelve dressings, field, large. 2. 36 dressings, field, medium. 3. 36 dressings, field, small. 4. Two boxes of bandage, gauze, 3 in., 12 per box. 5. Two boxes of band-

Code	Unit designation	Unit wt	Unit data	
			No. Pkgs	Contents
				per box.
INDIA UNIFORM	Bandage Unit	22 lbs	1	1. Six packages of gauze pads 2 x 2 in. 2. Four packages gauze pads 4 x 4 in. 3. Six packages bandages, absorbent, adhesive (bandaid) 100 per package. 4. Six spools of adhesive tape, 3 in. 5. 12 packages of cotton, sterile, 1 oz package. I 6. Two boxes of roller gauze, 2 in., 12 per box. 7. Two boxes of roller gauze, 3 In., 12 per box. 8. Two boxes of roller gauze, 4 in., 12 per box. 9. Two boxes of muslin bandage, 24 per box.
INDIA VICTOR	Elastic Bandage Unit	25 lbs	1	1. One box of bandages, cotton, elastic 2 in., 12 per box. 2. Two boxes of bandages, cotton, elastic 3 in., 12 per box. 3. Four boxes of bandages, cotton, elastic 4 in., 12 per box. 4. One box of bandages, cotton, elastic 6 in., 12 per box. 5. Two spools of adhesive tape, 3 in.
INDIA WHISKEY	Litter Set	46 lbs	1	1. Two litters, folding. 2. Two blankets.
INDIA XRAY	Mountain Litter Set	44 lbs	1	1. One litter, rigid, mountain. 2. Three blankets.

Code	Unit designation	Unit wt	Unit data	
			No Pkgs	Contents
INDIA YANKEE	Blanket Set	50 lbs	1	Ten blankets.
INDIA ZULU	Orthopedic Cast Set No. 1.	50 lbs	1	1. One roll stockinet, 3 in. x 25 yds. 2. One roll stockinet, 4 in. x 25 yds. 3. One roll stockinet, 6 in. x 12 yds. 4. One roll bandage, felt, 3 in. roll. 5. One roll bandage, felt, 4 in. roll. 6. One roll bandage, felt, 6 in. roll. 7. Two boxes of bandages, cotton, plaster impregnated 3 in., 12 per box. 8. One box bandages, cotton, plaster impregnated, 6 in., 12 per box. 9. Six heels, orthopedic cast. 10. One pair shears, plaster cast, hand. 11. One saw, plaster cast cutting, hand. 12. Two plastic buckets (non-standard).
JULIET ALPHA	Orthopedic Cast No. 2.	50 lbs	1	1. Three boxes bandages, cotton, plaster impregnated 3 in., 12 per box. 2. Three boxes bandages, cotton, plaster impregnated 4 in., 12 per box. 3. Three boxes bandages, cotton, plaster impregnated 6 in., 12 per box.
JULIET BRAVO	Eye, Ear, Nose, Throat Set.	25 lbs	1	1. One nasal speculum 2. One myringotome. 3. One box of cotton

Code	Unit designation	Unit wt	Unit data	
			No. Pkgs	Contents
				per box. 4. One box tongue depressors, 100 per box. 5. Two bottles Chlorpheniramine maleate (chlortrimeton), 40 mg, 1,000 per bottle. 6. Two boxes Tetrahydrozoline hydrochloride (tyzine) nose drops, 12 per box. 7. One bottle APC, tablets, 1,000 per bottle. 8. One bottle ASA, 5 gr. tablet, 1,000 per bottle. 9. Six bottles Codeine, ½ gr. tablet, 100 per bottle. 10. One box MBA eardrops, 12 per box. 11. Two bottles Terpin hydrate, 1 lb bottle. 12. One bottle Boric acid, 1 lb bottle. 13. Six tubes Polymixini B-Bacitracin-Neomycin ophthalmic ointment, ½ oz tube. 14. Three tubes cortisone opthalmic ointment, ½ oz tube. 15. One mirror, laryngeal.
JULIET CHARLIE	Skin Disease Unit	25 lbs	1	1. 24 tubes undecylenic acid ointment, 2 oz tube. 2. Twelve cans of foot powder. 3. Three boxes of bacitracin-neomycin ointment, ½ oz tube, 6 per box. 4. Four tubes of dibucaine ointment, 1 oz tube. 5. One jar of wool fat. 1 lb jar.

Code	Unit designation	Unit wt	Unit data	
			No. Pkgs	Contents
				6. One can of petrola- tum, 1 lb can. 7. One jar of zinc oxide, 1 lb jar. 8. Two bottles Burow's solution, tablets, 500 per bottle. 9. Eight bottles of po-tassium permanga-nate tablets, 5 grain, 100 per bottle. 10. One bottle of methylrosaniline chloride, 1 lb bottle. 11. Six bottles selenium sulfide suspension, four oz plastic bottle. 12. Twelve tubes of hydrocortisone acetate ointment, 1/6 oz tube. 13. Six tubes of benzene hexachloride, 1% ointment, 60 gm tube. 14. One bottle calamine, USP, 1 l bottle. 15. One can methylsali-cylate (oil of winter-green) 1 lb can.
JULIET DELTA	Depressant Stimulants Unit.	20 lbs	1	1. 24 bottles phenobar-bital, 30 mg tablets, 100 per bottle, 2. Two bottles secobar-bital, 100 mg capsule, 500 per bottle. 3. Two bottles mepro-bamate, 400 mg tablet, 500 per bottle. 4. Twelve bottles of dephenhydrainine hydrochloride (Benadryl) 50 mg capsule, 100 per bot-tle. 5. Twelve ampules of amobarbital sodium, sterile intravenous injection, 7% grains, ampule. 6. Two boxes pentylen-etetrazol, 100 mg,

Code	Unit designation	Unit wt	Unit data	
			No. Pkgs	Contents
				box. 7. Ten bottles of dextroamphetamine, 5.0 mg tablet, 100 per bottle.
JULIET ECHO	Gastrointestinal Diseases Unit.	20 lbs (-)	1	1. Two bottles of Kaolin, 1 lb bottle (powder). 2. One bottle of pectin, 1 lb bottle (powder). 3. Five boxes of aluminum hydroxide gel, and magnesium trisilicate tablet, 100 per box. 4. One bottle of tincture belladonna, 1 pint bottle. 5. Six bottles of cascara sagrada tablets, 100 per bottle. 6. Two bottles of opium, tincture, USP ¼ pint. 7. Six bottles propantheline, bromide, 15 mg, 100 per bottle. 8. Twelve bottles phenobarbital, 30 mg tablet, 100 per bottle. 9. Three bottles of chloramphenicol, 250 mg tablet, 100 per bottle. 10. Three bottles tetracycline, 250 mg tablet, 100 per bottle. 11. Six bottles of neomycin sulfate, 500 mg tablet, 20 per bottle. 12. Six bottles of chlor-promazine, 25 mg, 50 per bottle 13. Six bottles of Bismuth subcarbonate 5

Code	Unit designation	Unit wt	Unit data	
			No, Pkgs	Contents
				bottle. 14. Four tubes of dibucaine ointment, 1 oz tube
JULIET FOXTROT	Tropical Disease Unit	20 lbs		1. Six bottles of tetracycline, 250 mg tablets, 100 per bottle. 2. 48 bottles of hexylresorcinol, 200 mg, 5 per bottle. 3. 24 bottles of carbasone, 250 mg tablets, 20 per bottle. 4. Six boxes of stibophen, injection, 64 mg., 5 cc, 10 per box. 5. Twelve bottles of diethylcarbamazine, 50 mg tablet, 100 per bottle. 6. Six bottles of chloro-quin, 500 mg tablet, 100 per bottle. 7. Six bottles of primaquine, 25 mg tablet.
JULIET GOLF	Malnutrition Unit	25 lbs	1	1. 36 bottles of decavitamins, 100 per bottle. 2. One bottle of ferrous sulfate, 5 grain tablets, 1,000 per bottle. 3. Twelve bottles of menadione, 2.0 mg tablets, 50 per bottle. 4. Six bottles of ascorbic acid, 50 mg tab-lets, 100 per bottle. 5. Six bottles of riboflavin, 1.0 mg tablets, 100 per bottle. 6. Six bottles of Thiamin, 5.0 mg tablets, 100 per bottle.

Code	Unit designation	Unit wt	Unit data	
			No. Pkgs	Contents
				7. Four cans of skimmed milk, powdered, 1 lb can.
JULIET HOTEL	Malaria Unit.	20 lbs	1	1. 24 bottles of chloroquin, 500 mg, 100 per bottle. 2. Twelve bottles of primaquin, 25 mg, 100 per bottle. 3. Two bottles of acetylsalicylic acid, 5 grains, 100 per bottle.
JULIET INDIA	Antibiotics No. 1.	25 lbs	1	1. 50 packages of procaine penicillin, 400,-000 unit cartridge, 10 per package. 2. Six cartridge syringes. 3. Twelve bottles of tetracycline, 250 mg tablets, 100 per bottle. 4. Twelve bottles chloramphenicol, 250 mg tablets, 100 per bottle.
JULIET JULIET	Antibiotics No. 2.	18 lbs	1	1. Two bottles of sulfasoxizole, 500 mg tablets, 1,000 per bottle. 2. Twelve bottles of penicillin, 250,000 unit tablets, 100 per bottle.
JULIET KILO	Antibiotics No. 3.	20 lbs	1	1. One box of penicillin, crystalline, 200,-000 unit vial, 100 per box. 2. Three boxes of tetracycline hydrochloride, intravenous, 250 mg, 12 per box.

Code	Unit designation	Unit wt	Unit data	
			No Pkgs	Contents
				chloramphenicol, intravenous, 100 mg, 12 per box (non-standard item). 4. Two boxes of water for injection, sterile, 5.0 cc vials, 12 per box.
JULIET LIMA	Mass Immunization Set.	48 lbs	1	1. Four trays, instrument with cover. 2. One sterilizer, surgical instrument fuel heated. 3. Six hand towels. 4. Two bowls, plastic (non-standard item). 5. Six cakes of soap, surgical, 4 oz cake. 6. Six containers of detergent, surgical, 5 oz container. 7. Six bottles of benzethonium chloride tablets, 80 per bottle. 8. Two Foerster forceps. 9. Two cans of alcohol, pint can. 10. Six bottles of Benzalkonium chloride, 25 cc bottle. 11. Six packages of gauze pads, 2 x 2 in. 12. Two packages of cotton, purified, rolled. 13. 48 syringes, 5.00 cc. 14. Twelve syringes, 10.0 cc. 15. Four syringes, 20.0 16. Two books, record, ruled. 17. Six pencils. 18. One package of gauze pads, 4 x. 4 in.

Code	Unit designation	Unit wt	Unit data	
			No. Pkgs	Contents
				needles, gauge # 23, 12 per box. 20. Twelve boxes needles, gauge # 22, 12 per box. 21. Three boxes needles, gauge # 20, 12 per box. 22. Three boxes needles, gauge # 18, 12 per box.
JULIET MIKE	Diphtheria Immunization. *Notes 4 and 5*	20 lbs	1	Diphtheria-pertussis-tetanus vaccine, 300 amplues.
JULIET NOVEMBER	Tetanus Immunization. *Notes 4 and 5*	20 lbs	1	300 ampules of tetanus toxoid, 5 cc ampule.
JULIET OSCAR	Typhoid Immunization. *Notes 4 and 5*	18 lbs	1	50 vials of typhoid-paratyphoid vaccine, 50 cc vial.
JULIET PAPA	Smallpox Immunization *Notes 4 and 5*	11 lbs	1	150 boxes of smallpox vaccine, 10 per box.
JULIET QUEBEC	Typhus Immunization. *Notes 4 and 5*	30 lbs	1	150 vials of typhus vac-cine 20 cc vial.
JULIET ROMEO	Cholera Immunization. *Notes 4 and 5*	26 lbs	1	Cholera vaccine, 20 cc vial.
JULIET SIERRA	Poliomyelitis Immunization *Notes 4 and 5*	30 lbs	1	300 bottles of Poliomye-litis vaccine, 9.0 cc bottle.
JULIET TANGO	Yellow Fever Immunization *Notes 4 and 5*	11 lbs	1	75 ampules of yellow fever vaccine, 20-dose ampule.
JULIET UNIFORM	Plague Immunization. *Notes 4 and 5*	22 lbs	1	150 vials of plague vac-cine USP, 20 cc vial.
JULIET VICTOR	Rabies Kit. *Notes 4 and 5*	18 lbs	1	1. 20 packages of

Code	Unit designation	Unit wt	Unit data	
			No. Pkgs	Contents
				14-dose package. 2. 80 ampules of anti-rabies serum, 1,000 units ampule.
JULIET WHISKEY	Tetanus Antitoxin. *Notes 4 and 5*	11 lbs	1	1. 100 bottles of tetanus antitoxin, 1,500 units per bottle. 2. 20 bottles of tetanus antitoxin, 20,000 units per bottle.
JULIET XRAY	Gamma Globulin. *Notes 4 and 5*	10	1	Ten bottles of globulin, immune serum, 10 cc bottle.

NOTES:

1. The field surgery set may be augmented with:
   a. Medical dispensary set.
   b. Supplemental supply set.
2. Quartermaster items.
3. Albumin does not withstand freezing.
4. Immunizations for 1,f>00 personnel.
5. Vaccine requires refrigeration.
6. Many medical packages weigh considerably less than 60 lbs. The supply agency adds items which are in constant demand to fill out lighter medical packages. Examples are-blankets, extra bandages, and dressings.
7. Chloroquine is deleted from packages used in non-malaria areas.

Section IV. WEAPONS AND AMMUNITION

Code	Unit designation	Unit wt	Unit data	
			No. Pkgs	Contents
MIKE ALPHA	Automatic Rifle (3).	250 lbs	3	1. One rifle, automatic, cal. .30, M1918A2 (20 lbs). 2. Thirteen magazines, AR (6 lbs). 3. One belt, ammunition, AR (2 lbs). 4. 480 rds, cartridge, AP cal. .30, 20 rd cartons, packed in ammunition can M8 (2 cans, 32 lbs).

Code	Unit designation	Unit wt	Unit data	
			No. Pkgs	Contents
				accessory packet (2 lbs) *Note 1*.
			1	960 rds, cartridge, AP, cal. .30, 20 rd cartons, packed in ammunition can M8 (4 cans, 64 lbs).
MIKE BRAVO	Carbine (20).	240 lbs	4	1. Five carbines, cal. .30, M-2 (30 lbs). 2. Fifteen magazines, carbine, 30 rd capac-ity (4 lbs). 3. 800 rds cartridge, ball, carbine cal. .30 M-1, packed in am-munition can M6 (1 can, 25 lbs).
MIKE CHARLIE	Light Machine-gun (2).	484 lbs	2	1. One machine gun, cal. .30, M1919A6 w/sholder stock and bipod (37 lbs). 2. 275 rds, cartridge, linked, cal. .30 4AP-1TR, packed in am-munition box M1A1 (1 box, 22 lbs). 3. One spare parts and accessory packet (2 lbs). *Note 1*.
			8	550 rds, cartridge, linked, cal. .30, 4AP-1TR, packed in ammu-nition box M1A1 (2 boxes, 44 lbs).
MIKE DELTA	Mortar (1).	320 lbs	1	1. One mortar, 60 mm, complete with base plate, mou nt and 1 sight (46 lbs). 2. One base plate, MI (4.5 lbs). 3. One spare parts and accessory packet (2 lbs). *Note 1*.
			5	

Code	Unit designation	Unit wt	Unit data	
			No. Pkgs	Contents
				HE, 60mm mortar, M49A2 packed in individual containers (53 lbs).
MIKE ECHO	Pistol (12).	90 lbs	2	1. Six pistols, automatic, cal. .45 M1911 A1 (15 lbs). 2. Eighteen magazines, pistol, cal. .45 (5 lbs). 3. 800 rds, cartridge, ball, cal. .45 packed in ammunition box M5 (1 box, 29 lbs). 4. Six shoulder stocks, pistol (6 lbs).
MIKE FOXTROT	Recoilless Rifle (2).	406 lbs	2	1. One rifle, 57mm, recoilless, T15E13 or M18, complete for shoulder firing, including telescope sight M86C (45 lbs). 2. One cover, overall, M123 (3 lbs). 3. One spare parts and accessory packet (5 lbs). *Note 1*.
			6	Eight rds, cartridge, HEAT, 57mm RR, M307, packed in indi-vidual containers (50 lbs).
MIKE GOLF	Rocket Launcher (3).	330 lbs	3	1. One launcher, Rocket, 3.5 inch, M20A1 or M20A1B1 (14 lbs). 2. Four rds, -A-f, HEAT, 3.5 inch M28A2 packed in individual containers (38 lbs).
			3	Six rds, rocket, HEAT, 3.5 in., M28A2, packed in individual containers (57 lbs).

Code	Unit designation	Unit wt	Unit data	
			No. Pkgs	Contents
MIKE HOTEL	Sniper Rifle (6).	165 lbs	3	1. Two rifles, cal. .30, M1C, complete (23 lbs). 2. 480 rds, cartridge, AP, cal. .30, 8 rd clips in bandoleers, packed in ammunition can M-8 (2 cans, 32 lbs).
MIKE INDIA	Submachine Gun (9).	175 lbs	3	1. Three submachine guns, cal. .45, M3A1 (21 lbs). 2. Nine magazines, submachine gun, 30- rd capacity (7 lbs). 3. 600 rds, cartridge, ball, cal. .45 packed in ammunition box M5 (1 box, 29 lbs).
MIKE JULIET	General Unit. Note 2	2560 lbs	46	1. One automatic rifle unit, 3 ARS. 2. One carbine unit, 20 carbines. 3. One light machine-gun unit, 2 LMGs. 4. One mortar unit, 1 mortar. 5. One pistol unit, 12 pistols. 6. One recoilless rifle unit, 2 RRS. 7. One rocket launcher unit, 3 RLS. 8. One sniper rifle unit, 6 rifles. 9. One submachine gun unit, 9 SMGS.
NOVEMBER ALPHA	Carbine Ammunition No. 1 (6400 rds).	200 lbs	4	1600 rds, cartridge, car-bine, ball, cal. .30, 50 rd cartons, packed in ammunition can M6 (2 cans, 50 lbs).
NOVEMBER BRAVO	Carbine Ammunition	200 lbs	3	1600 rds, cartridge, car-bine, ball, cal.

Code	Unit designation	Unit wt	Unit data	
			No. Pkgs	Contents
	rds).			packed in ammunition can M6 (2 cans, 50 lbs).
		200 lbs	1	1. 800 rds, cartridge, carbine, ball, cal. .30, 50 rd cartons, packed in ammunition can M6 (25 lbs). 2 800 rds, cartridge, carbine, tracer, cal .30, 50 rd cartons, packed in ammunition can M6 (25 lbs).
NOVEMBER CHARLIE	Rifle Ammunition No. 1 (2880 rds).	192 lbs	3	960 rds, cartridge, AP, cal. .30, 20 rd cartons, packed in ammunition can M8 (4 cans, 64 lbs).
NOVEMBER DELTA	Rifle Ammunition No. 2 (2880 rds).	192 lbs	2	960 rds, cartridge, AP, cal. .30, 20 rd cartons, packed in ammunition can M8 (4 cans, 64).
			1	1. 240 rds, cartridge, AP, cal. .30, 20 rd cartons, packed in ammunition can M8 (1 can, 16 lbs). 2 720 rds, cartridge, tracer, cal. .30, 20 rd cartons, packed in ammunition can M8 (3 cans, 48 lbs).
NOVEMBER ECHO	Rifle Ammunition No. 3 (1920 rds).	128 lbs	2	960 rds, cartridge, AP, cal. .30, 8 rd clips in bandoleers packed in ammunition can M8 (4 cans, 64 lbs).
NOVEMBER FOXTROT	Machinegun Ammunition No. 1 (2200 rds).	176 lbs	4	550 rds, cartridge, linked, cal. .30, 4AP-1TR, packed in ammu-nition box Ml A1 (2 boxes, 44 lbs)

Code	Unit designation	Unit wt	Unit data	
			No. Pkgs	Contents
NOVEMBER GOLF	Machinegun Ammunition No. 2 (2200 rds).	176 lbs	4	550 rds, cartridge, linked, cal. .30, 2AP-2API-1TR packed in ammunition box MIA1 (2 boxes, 44 lbs).
NOVEMBER HOTEL	Pistol Ammunition (2400 rds).	232 lbs	4	1200 rds, cartridge, ball, cal. .45, 50 rd cartons, packed in ammunition can M5 (2 cans, 58 lbs).
NOVEMBER INDIA	Recoilless Rifle Ammunition No. 1 (48 rds).	300 lbs	6	Eight rds, cartridge, HEAT, 57mm RR, M307 packed in indi- vidual containers (50 lbs).
NOVEMBER JULIET	Recoilless Rifle Ammunition No. 2 (48 rds).	300 lbs	6	Eight rds, cartridge, smoke WP, 57mm RR, M308, packed in indi- vidual containers (50 lbs).
NOVEMBER KILO	Recoilless Rifle Ammunition No. 3 (48 rds).	300 lbs	6	Eight rds, cartridge, HE, 57mm RR, M306, packed in individual containers (50 lbs).
NOVEMBER LIMA	Mortar Ammunition No. 1 (60 rds).	121 lbs	4	Fifteen rds, shell, HE, 60mm mortar, M49A2, packed in individual containers (53 lbs).
NOVEMBER MIKE	Mortar Ammunition No. 2 (24 rds).	110 lbs	2	Twelve shell, smoke WP, 60mm mortar, M302 packed in individual containers (55 lbs).
NOVEMBER NOVEMBER	Mortar Ammunition No. 3 (24 rds).	96 lbs	2	Twelve shell, illuminat- ing, 60mm mortar packed in individual containers (55 lbs).
NOVEMBER OSCAR	Rocket Launcher Ammunition (48 rds).	456	8	Six rds, rocket, HEAT, 3.5 inch, M28A2 packed in individual containers (57 lbs).
NOVEMBER PAPA	Grenade No. 1 (50 rds).	60 lbs	1	50 grenades, hand, fragmentation, M26 (T38E1) packed in in- dividual container (60 lbs).

Code	Unit designation	Unit wt	Unit data	
			No. Pkgs	Contents
NOVEMBER QUEBEC	Grenade No. 2 (50 rds).	45 lbs	1	50 grenades, hand, illuminating, MK1, packed individual con- tainer (45 lbs).
NOVEMBER ROMEO	Pyrotechnic Signal No. 1 (60 rds).	15 lbs	1	1. Twenty signal, red star, ground, red star, parachute, M126(T72), packed in individual container (5 lbs). 2. Twenty signal, ground, white star, parachute, M127(T73), packed in individual container (5 lbs). 3. Twenty signal, ground, green star, cluster M127 (T71), packed in individual container (5 lbs). 4. One projector, pyrotechnic.
NOVEMBER TANGO	Pyrotechnic Signal No. 2 (60 rds).	15 lbs	1	1. Twenty signal, green smoke, parachute M128(T74), packed in individual container (5 lbs). 2. Twenty signal, red smoke, parachute M129(T75), packed in individual container (5 lbs). 3. Twenty signal, yellow smoke, streamer M139 (T76), packed in individual container (5 lbs). 4. One projector, pyrotechnic.

NOTES:

1. The spare parts and accessory packet includes items most subject to damage or wear and tools required for the care and maintenance of the weapon.
2. The general unit contains the basic weapons for a type guerrilla platoon plus weapons peculiar to weapons platoons.
3. General—
   a. Weapons units contain cleaning and preserving material such as rods, lubricants and patches.
   b. Ammunition is stripped of its outer shipping containers and delivered in its inner weatherproof container.

Section V. QUARTERMASTER

Code	Unit designation	Unit wt	Unit data	
			No. Pkgs	Contents
QUEBEC ALPHA	Clothing and Equipment 40 personnel. *Notes 1 and 2*	840 lbs	20	Two man unit consisting of— 1. Two belts, pistol OD. 2. Two blankets, OD. 3. Two pair boots, com-bat. 4. Two coats, man's, water resistant sateen (field jacket). 5. Two canteens, dis-mounted w/cup and cover. 6. Two caps, field, poplin. 7. Two ponchos, coated nylon, OG-107. 8. Two pouches and packets, first aid. 9. Two pair socks, wool. 10. Two pair suspend-ers, trousers, OD-107. 11. Two pair trousers, men's, cotton, water resistant sateen (field trousers) (42 lbs).
QUEBEC BRAVO	Clothing and Equipment 100 personnel *Notes 1 and 2*	2100 lbs	50	Consists of 50 two-man units.
ROMEO ALPHA	Rations, indig-enous Personnel—100 men. *Note 3*	1750 lbs	35	High fat content meat or canned fish/poultry, sugar, tobacco, salt, coffee or tea, grain flour or rice, accessory items and water purification tablets (50 lbs).

Code	Unit designation	Unit wt	Unit data	
			No Pkgs	Contents
ROMEO BRAVO	Rations, Indigenous Personnel—500 men. *Note 3*	85001bs	170	High fat content meat or canned fish, poultry, sugar, tobacco, salt, coffee or tea, grain, flour or rice, accessory items and water purification tablets (50 lbs).
ROMEO CHARLIE	Special Rations-96 men. *Note 4*	136 lbs	4	24 food packets, survival, (arctic or tropic) (34 lbs).
ROMEO DELTA	Special Rations-96 men.	192 lbs	8	12 individual combat meals (24 lbs).
ROMEO ECHO	Packet, barter. *Note 5*	500 lbs	10	50 lbs packages.

NOTES:

1. Items vary with the climatic zone and season. This package is based on the temperate zone for spring, summer, and fall seasons. For winter, add gloves and 1 extra blanket per individual.
2. Clothing sizes are issued as small, medium, and large. Clothing is matched to size of boots. Boot size is included in the message requesting the clothing package. The packaging agency dictates matching of boot and clothing sizes based upon experience factors applicable to the operational area.
3. Special rations for indigenous personnel are determined by the area of operations. Allotment is 15 lbs per individual per month.
4. The food packet varies with the climatic zone.
5. Contents to be determined by the area of operations.

Section VI. SIGNAL

Code	Unit designation	Unit wt	Unit data	
			No. Pkgs	Contents
UNIFORM ALPHA	Batteries No. 1.	48 lbs	1	6 BA 279/U for AN/PRC-10.
UNIFORM BRAVO	Batteries No. 2.	50 lbs	1	20 BA 270/U for AN,/PRC-6.
UNIFORM CHARLIE	Batteries No. 3.	53 lbs	1	1. 15 BA 317/& (15 lbs). 2. 100 BA 32 (25 lbs). 3. Five BA 1264/U (10 lbs). 4. Two BA 58/U (1 lbs). 5. Two 13A 261/U (2 lbs).

Code	Unit designation	Unit wt	Unit data	
			No. Pkgs	Contents
UNIFORM DELTA	Field Wire (1 mile).	56 lbs	1	1. One mile wire WD-1 in dispensers, MX 306-two dispensers, total (52 lbs). 2. One tool equipment set TE-33, (2 lbs). 3. Tape, friction, 2 rolls, (1 lb). 4. Tape, rubber, 1 roll (1 lb).
UNIFORM ECHO	Flashlights (20).	45 lbs	1	1. 20 Flashlights (15 lbs). 2. 120 Batteries, BA 30 (30 lbs).
UNIFORM FOXTROT	Power Unit UGP-12 (1).	60 lbs	1	1. One engine generator (15 lbs). 2. 5 gals gasoline (42 lbs). 3. One qt oil, SAE 10 or 30 (3 lbs).
UNIFORM GOLF	Radio Set AN/PRC-10 (1).	42 lbs	1	1. One AN/PRC-10 complete (18 lbs). 2. Three batteries BA 279/U (24 lbs).
UNIFORM HOTEL	Radio Set AN/PRC-6 (2). *Note* d	43 lbs	1	1. Two AN/PRC-6, complete (8 lbs). 2. 14 batteries, BA 270/U (35 lbs).
UNIFORM INDIA	Telephones (4).	42 lbs	1	1. Four telephones, battery powered (38 lbs). 2. 16 batteries, BA 30 (4 lbs).
UNIFORM JULIET	Switchboard (1).	1134 lbs	28	1. One switchboard, SB 22, complete (40 lbs). 2. Eight batteries, BA 30, (2 lbs).
UNIFORM KILO	Signal Equipment Battalion. (90 lbs). *Note* 1	42 lbs	1	1. Two flashlight units 2. Eight radio set units, AN/PR-6 (344 lbs).

Code	Unit designation	Unit wt	Unit data	
			Max Pkgs	Contents
				AN/PRC-10 (210 lbs). 4. 200 Batteries, BA 30 (50 lbs). 5. Seven battery units, BA 270/U (250 lbs). 6. Five battery units, BA 279/U (240 lbs).
UNIFORM LIMA	Signal Equipment Area Command, HQ and HQ Company.	1599 lbs	31	1. One flashlight unit. (45 lbs). 2. Two radio set units, AN/PRC-10 (84 lbs). 3. One switchboard unit, SB-22 (42 lbs). 4. Three telephone units (126 lbs). 5. 20 wire units (1120 lbs). 6. 150 batteries, BA-30 (38 lbs). 7. Three battery units, BA 279 (144 lbs).
UNIFORM MIKE	Radio Set AN/GRC-109 (1)	92 lbs	1	1. Radio Transmitter, RT-3 (9 lbs). 2. Radio receiver, RR-2 (10 lbs). 3. Power supply, RP-1 (25 lbs). 4. Operating snares and accessories (6 lbs).
			1	5. Generator, G-43/G, complete, (22 lbs). 6. Adapter, RA-2 (4 lbs). 7. 16 batteries, BA 317/U (16 lbs).
UNIFORM NOVEMBER	Radio Set AN/GRC-9 (1).	99 lbs	1	1. Receiver-transmitter, RT-77 (32 lbs). 2. 15 batteries, BA 317/U (15 lbs).
			1	3. Generator, G-43/G, complete (22 lbs).

Code	Unit designation	Unit wt	Unit data	
			No. Pkgs	Contents
				4. Antennas and antenna accessories (23 lbs). 5. Audio accessories (5 lbs). 6. Spare parts kit (2 lbs).
UNIFORM OSCAR	Telephones Sound Powered.	45 lbs	1	1. 3 Reel Equipment, CE-11 (15 lbs). 2. 3 spools DR-8 with 3/8 mi (30 lbs) WD-1/TT.
UNIFORM PAPA	Switchboard Emergency.	108 lbs	1	1. Wire WD-1/TT, one mile, 2 dispensers (52 lbs).
			1	2. Wire WD-1/TT, one mile, 2 dispensers (52 lbs). 3. Switchboard, 93/GT, 1 ea (4 lbs).
UNIFORM QUEBEC	Radiac Detector Set.	40 lbs	1	1. Twenty radiac detector chargers 1578/PD. 2. Twenty radiac detectors IM 93 U/D.

NOTES:

1. Appropriate sets of crystals packed with AN/PRC-6 radios to allow frequency changes.
2. General—Cold weather batteries substituted when appropriate to season and area.

Section VII. SPECIAL

Code	Unit designation	Unit wt	Unit data	
			No. Pkgs	Contents
XRAY ALPHA	River Crossing Unit No. 1.	50 lbs	1	1. Five life rafts, inflatable, one person capacity with CO2 cylinder and accessory kit. 2. Five life preservers, yoke with gas cylinder. 3. Five paddles, boat, five feet long.
XRAY BRAVO	River Crossing Unit No. 2.	90 lbs	2	1. One life raft, inflatable, seven person capacity, with CO2 cylinder and accessory kit. 2. Seven life preservers yoke with gas cylinders. 3. Four paddles, boat, five feet long.

APPENDIX III

AREA STUDY GUIDE

Section I. INTRODUCTION

4. General

This appendix is an area study outline for Special Forces personnel. Sections II and III are to be used for study of the region of expected wartime assignment and as a guide for a more detailed evaluation of a selected country. The outline provides a systematic consideration of the principal factors which influence Special Forces operational planning.

5. Purpose

The purpose of the area study guide is to provide a means for acquiring and retaining essential information to support operations. Although the basic outline is general in nature, it provides adequate coverage when time is limited. As more time is made available for study, various subjects should be divided and further subdivided to produce a more detailed analysis of the area.

6. Technique of Preparation

The maximum use of graphics and overlays is encouraged. Most of the subsections lend themselves to production in graphical or overlay form.

Section II. GENERAL AREA STUDY

7. General

a. Political.

(1) Government, international political orientation, and degree of popular support.

(2) Attitudes of identifiable segments of the population to ward the United States, its allies and the enemy.

(3) National historical background.

(4) Foreign dependence and/or alliances.

(5) National capitol and significant political, military and economic concentrations.

b. Geographic Positions.

(1) Areas and dimensions.

(2) Latitude and climate.

(3) Generalized physiography.

(4) Generalized land utilization.

(5) Strategic location.

(a) Neighboring countries and boundaries.

(b) Natural defenses including frontiers.

(c) Points of entry and strategic routes.

c. *Population.*

(1) Total and density.

(2) Breakdown into significant ethnic and religious groups.

(3) Division between urban, rural, and/or nomadic groups.

(a) Large cities and population centers.

(b) Rural settlement patterns.

(c) Areas and movement patterns of nomads.

d. National Economy.

(1) Technological standards.

(2) Natural resources and degree of self-sufficiency.

(3) Financial structure and dependence upon foreign aid.

(4) Agriculture and domestic food supply.

(5) Industry and level of production.

(6) Manufacture and demand for consumer goods.

(7) Foreign and domestic trade and facilities.

(8) Fuels and power.

(9) Telecommunications and radio systems.

(10) Transportation-U.S. standards and adequacy.

(a) Railroads.

(b) Highways.

(c) Waterways.

(d) Commercial air installations.

e. *National Security.*

(1) Center of political power and the organization for nation defense.

(2) Military forces (Army, Navy and Air Force): summary of order of battle.

(3) Internal security forces-summary of organization and strength.

(4) Paramilitary forces: summary of organization and strength.

8. Geography

a. Climate. General classification of the country as a whole with normal temperatures, rainfall, etc., and average seasonal variations.

b. Terrain. General classification of the country noting outstanding features, i.e., coasts, plains, deserts, mountains, hills and plateaus, rivers, lakes, etc.

c. Major Geographic Subdivisions. Divide the: country into its various definable subdivisions, each with generally predominant topographical characteristics, i.e., coastal plains, mountainous plateau, rolling, heavily forested hills, etc. For each subdivision use the following outline in a more specific analysis o1.' the basic geography

(1) *Temperature.* Variations front normal and, noting tire. months in which they may occur, any extremes that would affect operations.

(2) *Rainfall and Snow.* Same as c(1), above.

(3) *Wind and Visibility.* Same as c (l) , above.

(4) *Relief.*

(a) General direction of mountain ranges or ridge lines and whether hills and ridges are dissected.

(b) General degree of slope.

(c) Characteristics of valleys and plains.

(d) Natural routes for and natural obstacles to cross-country movement.

(5) *Land utilization.* Note any peculiarities, especially the following:

(a) Former heavily forested areas subjected to widespread cutting or dissected by paths and roads; also, the reverse, i.e., pasture or waste land which has been reforested.

(b) Former waste or pasture land that has been resettled and cultivated—now being fanned or the reverse (former rural countryside that has been depopulated and allowed to return to waste land).

(c) Former swamp or marsh land that has been drained; former desert or waste land now irrigated and cultivated; and lakes created by post-1945 darns.

(d) Whenever not coincidental with c(5) (a), (b), or (c), above, any considerable change in rural population density since 1945 is noted.

(6) *Drainage.* General pattern.

(a) Main rivers, direction of flow.

(b) Characteristics of rivers and streams such as current, banks, depths, type of bottom and obstacles, etc.

(c) Seasonal variation, such as dry beds and flash floods.

(d) Large lakes or areas of marry ponds and/or swamps, (potential LZs for amphibious aircraft).

(7) *Coast.* Examine primarily for infiltration, exfiltration and resupply points.

(a) Tides and waves: winds and current.

(b) Beach footing and covered exit routes.

(c) Quiet coves arid shallow inlets or estuaries.

(8) *Geological basics.* Types of soil and rock formations (include areas for potential Us for light aircraft.

(9) *Forests and Other Vegetation.* Natural or cultivated.

(a) Type, characteristics and significant variations from tire norm and at the different elevations.

(b) Cover or concealment-density, seasonal variation.

(10) *Water.* Ground, surface, seasonal and potable.

(11) *Subsistence.* Noting whether seasonal or year-round.

(a) Cultivated-vegetables, grains, fruits, nuts, etc.

(b) Natural-berries, fruits, nuts, herbs, etc.

(c) Wild life-animals, fish and fowl.

9. People

Tire following suboutline should be used for an analysis of the population in any given region or country or as the basis for art examination of the people within a subdivision as suggested in 5c. In all events particular attention should be given to those areas within a country where the local inhabitants have peculiarities and arc at considerable variance in one or more ways from the normal, national way of life.

a. Basic Racial Stock and Physical Characteristics.

(1) Types, features, dress and habits.

(2) Significant variations from the norm.

b. Standard of Living and Cultural (Education) Levels.

(1) Primarily note the extremes away from average.

(2) Class structure. (Degree of established social stratification and percentage of population in each class.)

c. Health and Medical Standards.

(1) Common Diseases.

(2) Standards of Public Health.

(3) Medical Facilities and Personnel.

(4) Potable water supply.

(5) Sufficiency of medical supplies and equipment.

d. Ethnic Components. This should be analyzed only if of sufficient size, strength and established bonds to constitute a dissident minority of some consequence.

(1) Location or concentration.

(2) Basis for discontent and motivation for change.

(3) Opposition to majority and/or to the political regime.

(4) Any external or foreign ties of significance.

e. Religion.

(1) Note wherein the national religion definitely shapes the actions and attitudes of the individual.

(2) Religious divisions. Major and minor religious groups of consequence. See d(1) through (4) above.

f. Traditions and Customs. (Particularly taboos.) Note where they are sufficiently strong and established that they may influence an individual's actions or attitude even during a tear situation.

g. Rural Countryside.

(1) Peculiar or different customs, dress and habits.

(2) Village and farm buildings—construction materials.

h. Political Parties or Factions.

(1) If formed around individual leaders or based on established organizations.

(2) If a single dominant party exists, is it nationalistic in origin or does it have foreign ties?

(3) Major legal parties with their policies and goals.

(4) Illegal or underground parties and their motivation.

(5) Violent opposition factions within major political organizations.

i. Dissidence. General active or passive potential, noting if dissidence is localized or related to external movements.

j. Resistance. (Identified movements.) Areas and nature of activities, strength, motivation, leadership, reliability, possible contacts and external direction or support.

k. Guerrilla Groups. Areas and nature of operations, strength, equipment, leaders reliability, contacts and external direction or support.

10. Enemy

a. Political

(1) *Outside power.* (Number and status of nonnational personnel, their influence, organization and mechanism of control.)

(2) *Dominant National Party.* Dependence upon and ties with an outside power; strength, organization, and apparatus; evidences of dissension at any level in the party; and the location of those areas within the country that are under an especially strong or weak nonnational control.

b. Conventional Military Forces. (Army, Navy, Air Force.)

(1) Nonnational or occupying forces in the country.

(a) Morale, discipline, and political reliability.

(b) Personnel, strength.

(c) Organization and basic deployment.

(d) Uniforms and unit designations.

(e) Ordinary and special insignia.

(f) Leadership (officer corps).

(g) Training and doctrine.

(h) Equipment and facilities.

(i) Logistics.

(j) Effectiveness (any unusual capabilities or weak nesses).

(2) National (indigenous) forces (Army, Navy, Air Force). See *(a)* through *(j)* above.

c. Internal Security Forces (including border guards).

(1) Strength and general organization, distinguishing between nonnational and national elements.

(a) Overall control mechanism.

(b) Special units and distinguishing insignia.

(c) Morale, discipline and relative loyalty of native personnel to the occupying or national regime.

(d) Nonnational surveillance and control over indigenous security forces.

(e) Vulnerabilities in the internal security system.

(2) Deployment and disposition of security elements.

(a) Exact location down to the smallest unit or post.

(b) Chain of command and communication.

(c) Equipment, transportation and degree of mobility.

(d) Tactics (seasonal and terrain variations).

(e) Methods of patrol, supply and reinforcements.

(3) The location of all known guardposts or expected wartime security coverage for all types of installations, particularly along main LOCs (railroads, highways, and telecommunication lines) and along electrical power and POL lines.

(4) Exact location and description of the physical arrangement and particularly of the security arrangements of all forced labor or concentration camps and any potential POW inclosures.

(5) All possible details, preferably by localities, of the types and effectiveness of internal security controls, including check points, identification cards, passports and travel permits.

11. Targets

The objective in target selection is to inflict maximum damage on the enemy with minimum expenditure of men and materiel. Initially, the operational capabilities of a guerrilla force may be limited in the interdiction or destruction of enemy targets. The target area and the specific points of attack must be studied, carefully planned and priorities established. In general, targets are listed in order of priority.

a. Railroads.

(1) Considerations in the selection of a particular line

(a) Importance, both locally and generally.

(b) Bypass possibilities.

(c) Number of tracks and electrification.

(2) Location of maintenance crews, reserve repair rails and equipment.

(3) Type of signal and switch equipment.

(4) Vulnerable points.

(a) Unguarded small bridges or culverts.

(b) Cuts, fills, overhanging cliffs or undercutting streams.

(c) Key junctions or switching points.

(d) Tunnels.

(5) Security system.

b. Telecommunications.

c. POL.

d. Electric Power.

e. Military Storage and Supply.

f. Military Headquarters and Installations.

g. Radar and Electronic Devices.

h. Highways.

i. Inland Waterways-Canals.

j. Seaports

k. Natural and synthetic gas lines.

l. Industrial plants.

Note. Targets listed in b through l are divided a subsections generally as shown in a above. Differences in subsections are based upon the peculiarities of the particular target system.

Section III. OPERATIONAL AREA INTELLIGENCE

12. General

This is a guide for operational area intelligence. The attached outline serves to bring the essentials into focus. It is built upon section 11, General Area Study Guide, but narrows the factors so that they apply to a relatively small and specific area. It refines the critical elements and puts them into the perspective of an actual operation at a given time.

13. Purpose

To outline the development of detailed intelligence on an assigned guerrilla warfare operational area to support the commitment of a special forces detachment.

14. Format

a. Select those elements that are applicable to the situation and the assigned guerrilla warfare operational area for the time of the year from section II. Use appropriate sections of paragraphs 5-8.

b. Cull all nonessentials and prepare a straightforward summation of basic facts.

c. Note serious gaps in data as processed in b above and take immediate action to fill them with the most current reliable information.

d. Prepare or request graphics; large-scale sheets and special maps covering the assigned area; the latest photography and illustration or information sheets on targets within the area; town plans, sketches of installations, air and hydrographic charts related to the area.

e. Within the time limits permitted, assemble the material for ready reference. Then proceed to plot on maps and/or overlays, wherever feasible, the following:

(1) Recommended initial guerrilla bases and alternate bases.

(2) Primary and alternate DZs, LZs, or points for other forms of infiltration.

(3) Possible direction and orientation points for infiltration vehicles (aircraft, boat), landmarks, etc.

(4) Routes from infiltration point to likely guerrilla base with stopover sites.

(5) Points for arranged or anticipated contacts with friendly elements.

(6) Enemy forces known or anticipated-location, strength and capabilities.

(7) Estimate of enemy operations or movements during the infiltration period.

(8) Settlements and/or scattered farms in the vicinity of the infiltration point and tentative guerrilla bases.

(9) All railroads, highways, telecommunications, etc., in the guerrilla warfare operational area.

(10) All important installations and facilities.

(11) Significant terrain features.

(12) Off-road routes and conditions for movement in all directions.

(13) Distances between key points.

(14) Recommended point of attack on assigned target sys tems and selection of other potential target areas.

f. As time permits, continue to collect information and revise estimates in keeping with more current intelligence. Develop increasing detail on (1) through (14) above with special emphasis

(1) On the local indigenous inhabitants

(a) Ethnic origins and religion.

(b) Local traditions, customs and dress.

(c) Food, rationing, currency, etc.

(d) Attitudes toward the regime, the United States, for or against existing political ideologies.

(e) Any peculiarities, or variances among individuals or small groups.

(2) Enemy, military forces and installations.

(3) Internal security forces and police.

(a) Organization, locations and strengths.

(b) Unit designations, insignia and uniforms.

(c) Areas covered and unit responsibilities.

(d) Check points, controls and current documentation.

(e) Patrols and mobile units.

(4) Geographic features in greater detail.

(5) Approaching seasonal climatic changes and their effect upon weather and terrain.

(6) Target categories and target areas in greater detail.

APPENDIX IV

AREA ASSESSMENT

Section I. GENERAL AND INITIAL ASSESSMENT

15. General

a. In order to plan and direct operations, Special Forces detachment commanders need certain basic information about the operational area. This information, when gathered or confirmed in the operational area, is called an area assessment,

b. An area assessment is the immediate and continuing collection of information started after infiltration in a guerrilla warfare operational area. It has the following characteristics:

(1) It confirms, corrects, or refutes previous intelligence of the area acquired as a result of area studies and other sources prior to infiltration.

(2) It is a continuing process.

(3) It forms the basis for operational and logistical planning for the area.

(4) In addition to information of the enemy, weather, and terrain, it needs information on the differently motivated segments of the civil population and the area of operations.

c. The information developed as a result of the area assessment should be transmitted to the SFOB only when there is sufficient deviation from previous intelligence and the information would have an impact on the plans of higher headquarters. The SFOB prescribes in appropriate SOP's and annexes those items to be reported.

d. The following outline, containing the major items of interest to the area command, is an example of how such an assessment may be accomplished.

e. Emphasis and priority on specific items fluctuates with the situation.

f. This outline is designed to facilitate the collection processing, and collation of the required material and may be considered to have two degrees of urgency.

(1) Immediate. Initial assessment includes those items deemed essential to the operational detachment immediately following infiltration. These requirements must be satisfied as soon as possible after the detachment arrives in the operational area.

(2) Subsequent. Principal assessment, a continuous operation, includes those collection efforts which support the continued planning and conduct of operations. It forms the basis for all of the detachment's subsequent activities in the operational area.

16. Initial Assessment

a. Location and orientation.

b. Detachment physical condition.

c. Overall security,

(1) Immediate area.

(2) Attitude of the local population.

(3) Local enemy situation.

d. Status of the local resistance elements.

Section II. PRINCIPAL ASSESSMENT

17. The Enemy

a. Disposition.

b. Composition, identification, and strength.

c. Organization, armament, and equipment.

d. Degree of training, morale, and combat effectiveness.

e. Operations.

(1) Recent and current activities of the unit.

(2) Counter guerrilla activities and capabilities with particular attention to: reconnaissance units, special troops (airborne, mountain, ranger type), rotary wing or vertical lift aviation units, counterintelligence units, and units having a mass CBR delivery capability.

f. Unit areas of responsibility.

g. Daily routine of the units.

h. Logistical support to include

(1) Installations and facilities.

(2) Supply routes.

(3) Method of troop movement.

i. Past and current reprisal actions.

18. Security and Police Units

a. Dependability and reliability to the existing regime and/or the occupying power.

b. Disposition.

c. Composition, identification, and strength.

d. Organization, armament, and equipment.

e. Degree of training, morale, and efficiency.

f. Utilization and effectiveness of informers.

g. Influence on and relations with the local population.

h. Security measures over public utilities and government installations.

19. Civil Government

a. Controls and restrictions, such as:

(1) Documentation.

(2) Rationing.

(3) Travel and movement restrictions.

(4) Blackouts and curfews.

b. Current value of money, wage scales.

c. The extent and effect of the black market.

d. Political restrictions.

e. Religious restrictions.

f. The control and operation of industry, utilities, agriculture, and transportation.

20. Civilian Population

a. Attitudes toward the existing regime and/or occupying power.

b. Attitudes toward the resistance movement.

c. Reaction to United States support of the resistance.

d. Reaction to enemy activities within the country and, specifically, that portion which is included in guerrilla warfare operational areas.

e. General health and well-being.

21. Potential Targets

a. Railroads.

b. Telecommunications.

c. POL.

d. Electric power..

e. Military storage and supply.

f. Military headquarters and installations.

g. Radar and electronic devices.

h. Highways.

i. Inland waterways and canals.

j. Seaports.

k. Natural and synthetic gas lines.

l. Industrial plants.

m. Key personalities.

22. Weather

a. Precipitation, cloud cover, temperature and visibility, seasonal changes.

b. Wind speed and direction.

c. Light data (BMNT, EENT, sunrise, sunset, moonrise, and moonset.)

23. Terrain

a. Location of areas suitable for guerrilla bases, units, and other installations.

b. Potential landing zones, drop zones and other reception sites.

c. Routes suitable for

(1) Guerrillas.

(2) Enemy forces.

d. Barriers to movement.

e. The seasonal effect of the weather on terrain and visibility.

24. Resistance Movement

a. *Guerrillas.*

(1) Disposition, strength, and composition.

(2) Organization, armament, and equipment.

(3) Status of training, morale, and combat effectiveness.

(4) Operations to date.

(5) Cooperation and coordination between various existing groups.

(6) General attitude towards the United States, the enemy and various elements of the civilian population.

(7) Motivation of the various groups.

(8) Caliber of senior and subordinate leadership.

(9) Health of the guerrillas.

b. *Auxiliaries and/or the Underground.*

(1) Disposition, strength, and degree of organization.

(2) Morale, general effectiveness and type of support.

(3) Motivation and reliability.

(4) Responsiveness to guerrilla and/or resistance leaders.

(5) General attitude towards the United States, the enemy, and various guerrilla groups.

25. Logistics Capability of the Area

a. Availability of food stocks and water to include any restrictions for reasons of health.

b. Agriculture capability.

c. Type and availability of transportation of all categories.

d. Types and location of civilian services available for manufacture and repair of equipment and clothing.

e. Supplies locally available to include type and amount.

f. Medical facilities to include personnel, medical supplies, and equipment.

g. Enemy supply sources accessible to the resistance.

APPENDIX V
GLOSSARY OF TERMS

26. General

a. This glossary of terms is provided to enable the user to have readily available terms unique to unconventional warfare found in this manual. Although some terms are contained in JCS Pub 1 and AR 320-5 they are reproduced here for the benefit of personnel not having ready access to those publications. Other terms are not found elsewhere but are in common image in Special Forces units and are more descriptive than other presently accepted terms or fill a gap in the absence of a term. Where differences exist between army terms and JCS terms, the JCS term is used because of its joint acceptance.

27. Terms

a. Area Command. The organization composed of Special Forces and resistance elements (guerrilla forces, auxiliaries and the underground) located within a guerrilla warfare operational area for the purpose of directing all area operations. Also called sector command when a subdivision of an area command. See unconventional warfare forces.

b. Area, Complex. An area complex consists of guerrilla bases and various supporting facilities and elements. The activities normally included in the area complex are: security and intelligence systems, communications systems, mission support sites, reception sites, supply installations, training areas, and other supporting facilities.

c. Auxiliary Force. That element of the area command established to provide for an organized civilian support of the resistance movement

d. Denied Area. Comprises the enemy homeland, enemy occupied territory and other areas in which the government or people are subject to the direct or indirect control of the enemy. By virtue of this enemy control, these areas are normally denied to friendly forces.

e. Evasion and Escape. That part of unconventional warfare whereby friendly military personnel and other selected individuals are enabled to emerge from enemy-held or unfriendly areas to areas under friendly control (JCS Pub 1).

f. Guerrilla. An armed combatant who engages in guerrilla warfare. A guerrilla belongs to a unit organized along military lines and may or may not be a member of a military force.

g. Guerrilla Base. A guerrilla base is a temporary site where installations, headquarters and units are located. There is usually more than one guerrilla base within an area complex. From a base, lines of communications stretch out connecting other bases and various elements of the area complex. Installations normally found at a guerrilla base are: command posts, training and bivouac areas, supply caches, communications and medical facilities. In spite of the impression of permanence of tile installations, a guerrilla base is considered temporary and tenant guerrilla units must be able to rapidly abandon the base when required.

h. Guerrilla Force. The overt, militarily organized element of the area command.

i. Guerrilla Warfare. Combat operations conducted in enemy held territory by predominantly indigenous forces on a military or paramilitary basis, to reduce the combat effectiveness, industrial capacity and morale of the enemy (AR 320-5).

j. Guerrilla Warfare Operational Area (Guerrilla Warfare Area, Operational Area). A geographical area in which the organization, development, conduct and supervision of guerrilla warfare and associated activities by special forces detachments assists the accomplishment of the theater mission. The terms operational area and guerrilla warfare area are used synonymously.

k. Guerrilla Warfare Operational Sector. A subdivision of a guerrilla warfare operational area within which a single special forces detachment is responsible for the organization, development, conduct, and supervision of guerrilla warfare and associated activities.

l. Joint Unconventional Warfare Task Force (JUWTF). An organization composed of elements of two or more services which is constituted and designated by the commander of a unified or specified command to plan for and direct unconventional warfare.

m. Mission Support Site. A relatively secure site, utilized by a guerrilla force as a temporary stopover point. It adds reach to guerrilla operations by enabling units to stay away from and go farther from bases for a longer period of time. Food, ammunition and the latest intelligence information may be made available at this site.

n. Special Forces Operational Base (SFOB).

(1) An organization which is composed of a special forces group and attached or supporting units to provide command, administration, training, operational supervision, logistical support and intelligence for committed special forces detachments.

(2) The location of the Special Forces group during operations.

o. Sponsoring Power. Any nation which supports a resistance effort

p. Subversion Against a Hostile State (Resistance). That part of unconventional warfare comprising actions by underground resistance groups for the purpose of reducing the military, economic, psychological, or political potential of an enemy. As resistance groups develop strength, their actions may become overt and their status shift to that of a guerrilla force (JCS Pub 1).

q. Unconventional Warfare. The three interrelated fields of guerrilla warfare, evasion and escape, and subversion. (JCS Pub 1).

r. Unconventional Warfare Forces. Forces who engage in unconventional warfare. For the purpose of this manual, UW forces include both U.S. forces (special forces detachments) and the sponsored resistance force (guerrillas, auxiliaries and the underground). Often used interchangeably with area command.

INDEX

BY ORDER OF THE SECRETARY OF THE ARMY

G. H. DECKER,
General, United States Army,
Chief of Staff.

Official:

R. V. LEE,
Major General, United States Army,
The Adjutant General.

Distribution:

Active Army:
DCSPER (2)
ACSI (2)
DCSOPS (10)
DCSLOG (2)
ACSRC (2)
CRD (1)
COA (1)
CINFO (1)
TIG (1)
TJAG (1)
TPMG (1)
Tech Stf, DA (1)
USCONARC (10)
ARADCOM (2)
ARADCOM Rgn (1)
OS Maj Comd (5)
LOGCOMD (1)
Armies (25)
Corps (3)
Div (2) except
Armor Div (5)
Div Arty (2)
Bde (1)
Regt/Gp (1) except
TOE 33105 (20)
Inf Bg (1)
Bn (1)
Co/Btry (1)
Br Svc Sch (5) except
USAIS (2530)
USASWS (1500)
USACGSC (5550)
USAINTC (415)

NG: State AG (3); Units organized under following TOE: 33106, 33107(1).

USAR: Same as Active Army.

For explanation of abbreviations used, see AR 32050.

* U. S. GOVERNMENT PRINTING OFFICE: 1961610531/13028

Made in the USA
Monee, IL
03 February 2026

43358971R00144